T0339407

Marine Systems Identification, Modeling, and Control

Marine Systems Identification, Modeling, and Control

Tony Roskilly

Rikard Mikalsen

AMSTERDAM • BOSTON • HEIDELBERG • LONDON
NEW YORK • OXFORD • PARIS • SAN DIEGO
SAN FRANCISCO • SYDNEY • TOKYO

Butterworth-Heinemann is an imprint of Elsevier

Butterworth Heinemann is an imprint of Elsevier
The Boulevard, Langford Lane, Kidlington, Oxford OX5 1GB, UK
225 Wyman Street, Waltham, MA 02451, USA
125 London Wall, London EC2Y 5AS, UK
525 B Street, Suite 1800, San Diego, CA 92101, USA

Notices
Knowledge and best practice in this field are constantly changing. As new research and experience broaden our understanding, changes in research methods, professional practices, or medical treatment may become necessary.

Practitioners and researchers must always rely on their own experience and knowledge in evaluating and using any information, methods, compounds, or experiments described herein. In using such information or methods they should be mindful of their own safety and the safety of others, including parties for whom they have a professional responsibility.

To the fullest extent of the law, neither the Publisher nor the authors, contributors, or editors, assume any liability for any injury and/or damage to persons or property as a matter of products liability, negligence or otherwise, or from any use or operation of any methods, products, instructions, or ideas contained in the material herein.

Library of Congress Cataloging-in-Publication Data
A catalog record for this book is available from the Library of Congress.

British Library Cataloguing-in-Publication Data
A catalogue record for this book is available from the British Library.

ISBN: 978-0-08-099996-8

For information on all Elsevier publications
visit our website at http://store.elsevier.com/

Printed and bound in the United Kingdom

Publisher: Joe Hayton
Acquisition Editor: Hayley Gray
Editorial Project Manager: Carrie Bolger
Production Project Manager: Susan Li
Designer: Inês Maria Cruz

Working together
to grow libraries in
developing countries

www.elsevier.com • www.bookaid.org

Preface

This book has evolved over many years of teaching dynamics and control to undergraduate and postgraduate students following marine technology programs. A key realization by the authors was that although most marine technology degree courses teach an introduction to dynamic systems and control theory, for the majority of students this will be the only engagement they have with this subject. Nevertheless, the teaching and learning materials available and used in such courses are usually written to prepare the student for further advanced studies in the subject, and very often not described in a marine context.

The book therefore aims to explain key topics while referring to marine-related systems and components and cover the dynamics and control in a practical manner, while keeping the theoretical and mathematical level which is expected in a degree-level course. Having extensive experience with distance-learning courses, one particular objective has been to make the text suitable for independent self-study. For this reason the authors demonstrate theory and system analysis using examples written not only in Matlab, which is the industrial standard simulation software for control engineering, but also in Scilab which is open source and freely available.

Contents

Chapter | One

Introduction

CHAPTER POINTS

- Introduction to control systems.
- History and background.
- Open- and closed-loop systems.
- System dynamics.
- Software tools.

1.1 INTRODUCTION TO CONTROL SYSTEMS

Control engineering is the science of altering the behavior of a dynamic process in a beneficial way. By dynamic process, we mean a process whose output(s) change as a continuous, time-varying function of the input(s). A simple example is the temperature of a room controlled by a boiler and radiator. The input to the system is the desired room temperature and the manner in which the actual room temperature responds is a dynamic function dependent on the physical parameters associated with the boiler, room, and the external conditions.

Control systems are key components in a range of industrial areas and applications, including marine and mechanical engineering, industrial manufacturing, chemical and process engineering, aviation, space flight, and electrical systems. Control systems also surround us in everyday life, in applications such as thermostats (in, e.g., hot water tanks and refrigerators), washing machines, consumer electronics, traffic lights, car cruise controls, etc.

The use of a control system in a mechanism, device, or process may have many different objectives. Of highest importance is to ensure *system stability*, for example controlling the reaction rate in a chemical process to prevent it going out of control. Control systems may also be used to optimize the operation of a plant, such as the adjusting of fuel and air flow rates to an engine to maximize fuel efficiency and minimize emissions. The use of active roll stabilization in ships modifies the dynamic behavior of the vessel in order to improve passenger comfort. Control systems may also be used to improve inherent performance limitations of a system, for example, to speed up the dynamic response. An example of the latter is the control of modern fighter jets,

Marine Systems Identification, Modeling, and Control. http://dx.doi.org/10.1016/B978-0-08-099996-8.00001-X

where advanced control is used to stabilize the aircraft and provide improved high-speed manoeuvrability.

In addition to a controller, control systems also usually include other elements:

- *Sensors* are used to measure some physical property of the process or plant and thus provide information for the controller, e.g., a thermo-couple device to provide temperature data or an optical encoder to measure the rotation of a motor.
- *Actuators* transduce signals from the controller to provide the input to and cause a change in behavior of the process or plant, e.g., a hydraulic ram to rotate the rudder in a steering gear system, a heater, or an electric motor transmission to drive a link of a robotic arm.

It is important to keep in mind that these components may also influence the behavior of the overall control system, for example, if they have a slow response or provide inaccurate or noisy data.

1.2 HISTORY OF CONTROL ENGINEERING

The field of control engineering has developed alongside the industrial revolution. The steam engine centrifugal governor, used by among others James Watt and illustrated in Figure 1.1, is often considered to be one of the first automatic controllers. The governor adjusts the steam supply to the engine according to the engine speed, ensuring stable operation under varying load conditions. (If the speed drops due to a load increase, the steam supply is increased, and vice versa.) These early control systems were developed by engineers in the mid-nineteenth century and were followed by a more formal mathematical analysis of dynamic systems by scientists such as James C. Maxwell and Oliver Heaviside.

What is currently considered to be standard techniques in feedback control systems design, including proportional-integral-derivative (PID) feedback control, root locus, and frequency response methods, were developed around

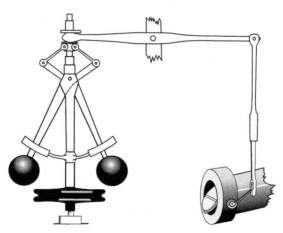

FIGURE 1.1 Centrifugal engine governor.

1920-1950 by engineers and scientists such as Walter R. Evans, Hendrik W. Bode, and Harry Nyquist. Evans developed the root locus method, a graphical method to determine the behavior of a system for variations in some design parameter, such as controller gain. Bode and Nyquist developed techniques to study the time-domain stability and behavior of a system based on its frequency-domain characteristics. We will use these tools later.

1.3 CONTROL SYSTEM STRUCTURE

Control systems can, in general, be classified as either open loop or closed loop. Let us look at what these terms mean, and what the structure of a control system looks like.

1.3.1 Open-loop systems

In open-loop systems, the output of a plant or process may be controlled by varying the input, but the actual output has no influence upon that input. This can be illustrated with a simple block diagram as shown in Figure 1.2.

An example of an open-loop control system is a room with a simple electric fire, illustrated in Figure 1.3. In this case, given an input (the mains supply switched on), the output (the room temperature) will eventually arrive at some constant level, i.e., the room temperature will become constant with time. The value of the output is dependent on the prevailing conditions or plant behavior, in this case the size of the heater and the heat losses from the room. If the rate of

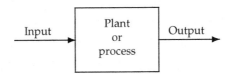

FIGURE 1.2 Open-loop control system.

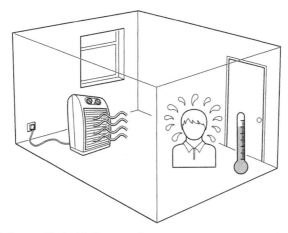

FIGURE 1.3 Room with electric fire example.

heat loss changes, for example, due to a change in outside temperature (external disturbance) or if someone opens a window (change in the plant), the room temperature will eventually settle at a new and different steady-state value. The open-loop control system does not know about the change in temperature and is therefore unable to correct for these external disturbances; when switched on the electric fire output is constant for all cases.

1.3.2 Closed-loop systems

In order for the system to be able to respond to external disturbances, variations in the input, or plant changes, and to achieve a desired output value, a *closed loop* (also known as feedback) system is required. Using a closed-loop control system, the actual output is continuously compared with the desired output value, in order that the controller can compensate for these system changes.

Consider the electric fire example from above: in a well-insulated room the temperature might rise to an uncomfortable level and the fire would have been switched off by someone sensing the room temperature. This situation is illustrated in Figure 1.4. Based on the "sensor signal" (the felt temperature), the person determines the action to be taken, i.e., acts as a controller, and switches the supply on or off. The actuator in this system is the heater, which is the component providing the heat energy input to the room. The controller adjusts the heater state, thereby modifying the flow of heat energy to the room. This type of control system is known as a closed-loop control system, or sometimes an error-actuated feedback control system.

1.3.3 System structure

In all closed-loop control systems, the output is measured and compared to the *desired* output (the setpoint). Any difference between the two (called an "error

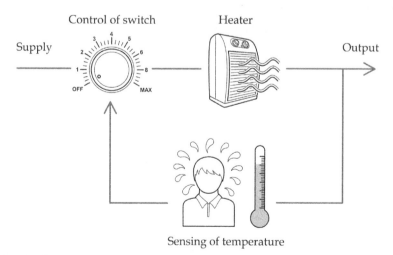

FIGURE 1.4 Room with electric fire and "manual" control.

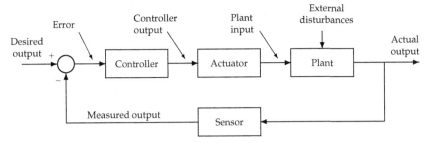

FIGURE 1.5 Feedback control system.

signal") forms the input to the controller. Open-loop control systems do not have this feedback loop, and can therefore not compensate for errors resulting from changes in the plant, variations in the setpoint, or disturbances acting on the system in the way that closed-loop systems can.

The standard block diagram representation of a feedback control system is shown in Figure 1.5. In many cases, the actuator and the plant are represented by one block, since they are often closely integrated.

In the block diagram, the blocks represent components of the system, whereas the connecting vectors represent some system state or signal. In the room temperature example, if controlled by a thermostat, the room is the plant, with the output being the physical room temperature and the setpoint is an electronic signal representing the desired output temperature. The sensor, a component of the system, converts the actual temperature to a measured temperature signal. The difference between the setpoint and the actual output value, the error signal, is fed to the controller. The controller output signal operates the actuator, the heater. The output from the actuator, which is the input to the plant, is the physical variable that we influence to achieve the desired output. (In this case, the heat flow to the room.)

In most modern control systems, the setpoint, sensor, error, and controller output signals are electronic signals. This is, however, not always the case; for example, the centrifugal governor discussed above is a purely mechanical controller. Such systems can also be fitted into the general structure above.

1.3.4 Marine control system examples

Control systems play an essential part in modern marine systems. There are numerous examples of maritime control systems; we will look at several of these later. A couple of introductory examples are included here.

Tank level control

Figure 1.6 illustrates a day tank where the supply line flow is automatically controlled with a valve to maintained a suitable tank level. The level sensor provides a measurement of the actual tank level to the controller, which again controls the fuel supply valve to maintain a desired tank level. Through the level

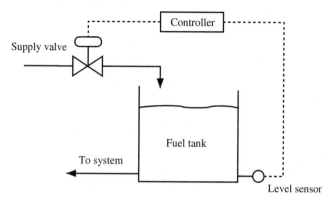

FIGURE 1.6 Fuel tank level control.

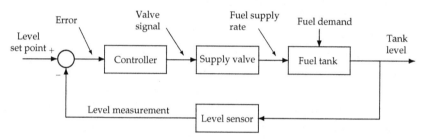

FIGURE 1.7 Tank level control block diagram.

measurement, the controller can respond to changes in fuel demand, hence this is a closed-loop control system.

Fitting this into the standard block diagram feedback control-loop structure, we get a system as shown in Figure 1.7. The level setpoint is provided in the controller, either programmed in or as an input that can be varied by an operator. The main external disturbance in this case are variations in the fuel demand, i.e., the output flow from the tank, which is what the control system should compensate for.

Fuel viscosity control

Controlling the viscosity of the fuel supplied to the propulsion engine(s) is essential and in order to achieve this, temperature control by electric or steam heating is typically used.

Figure 1.8 shows a common setup. The fuel is supplied through a fuel heater, through which a steam flow is controlled in order to provide the appropriate heating power. A feedback signal is provided by a viscosity meter downstream of the heater, which provides a information to the controller about the actual viscosity, in order that the opening of the steam supply valve can be regulated to produce the desired viscosity. Figure 1.9 shows the feedback system block diagram of this system.

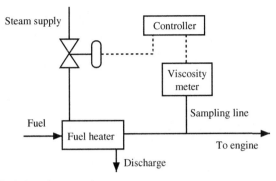

FIGURE 1.8 Fuel viscosity control.

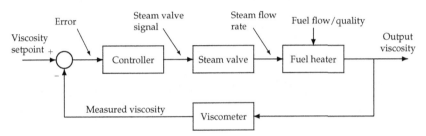

FIGURE 1.9 Fuel viscosity control block diagram.

Fin roll stabilization

Fin roll stabilization, as shown in Figure 1.10, is commonly used to improve the behavior of a ship under the impact of wave loads. By measuring the actual roll angle, a righting moment can be produced by controlling the angle of the fins.

Figure 1.11 shows the feedback block diagram representing this system. In this case, the controller demands a certain fin angle output, whereas the fin system, which could even be a feedback control system in its own right, will respond and produce a righting moment on the vessel.

1.4 SYSTEM DYNAMICS

A key concept in control engineering is system dynamics. The dynamics of the system determines how quickly it responds to controller outputs and to external disturbances. For the control system designer, the plant or process is often given; e.g., the design of a ship or a room of a given size. However, one may well be able to choose the actuator(s) and/or sensor(s) used in the control system.

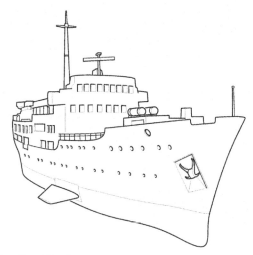

FIGURE 1.10 Fin roll stabilization.

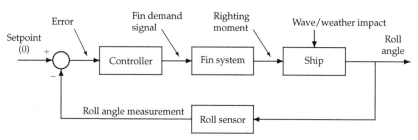

FIGURE 1.11 Fin roll stabilization control block diagram.

Choosing an appropriate actuator is important. If a ship steering gear does not have a powerful enough actuator it will only be able to move the rudder slowly under rough conditions. In the case of temperature control, if a boiler cannot supply sufficient heat, the room will remain cold for a long period of time; the system will be "sluggish."

Often we apply a control system to a process or plant to speed up the response time, although ultimately the process limits the performance. This is an important point to bear in mind; for example, we could easily over-specify a control system for a slowly changing process, spending a lot more money for a fast-acting control system that does not provide any gain in speed or accuracy of response.

1.4.1 Dynamic response examples

Consider the fuel tank level control example above; probably the simplest controller we can use on this plant is one that switches on the supply if the

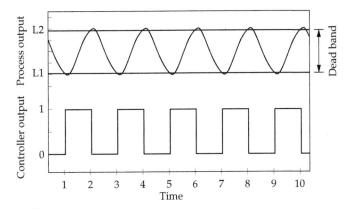

FIGURE 1.12 Controller output and plant response for on/off control.

tank level falls below a predetermined value, and switches the supply off when the level reaches a higher, second value. This is know as on/off, or bang-bang, control.

The response may look something like that shown in Figure 1.12. The two levels L1 and L2 are the setpoints for switching on or off the supply. Hence, the tank level will fluctuate between these levels (known as the dead band), and the supply will be either fully on or switched off. This system may be satisfactory for the tank control, where the main objective probably is to avoid the tank going empty, but for many systems, e.g., the viscosity controller and roll stabilization above, an oscillating response like this may not be acceptable.

If we consider the viscosity controller example from Figures 1.8 and 1.9, the control objective will be to maintain close to constant output viscosity (at some setpoint value) even in the case of varying fuel input conditions. A critical event in this system may be the change of fuel supply tank, after which a fuel with a different quality and temperature may be supplied.

Figure 1.13 illustrates some possible system responses following from such an event. We assume that at time zero the system is in a stable condition and the input conditions suddenly change. (Hence, our dynamic problem starts at $t = 0$.)

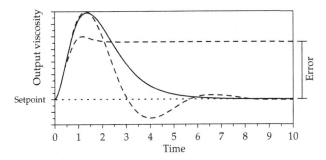

FIGURE 1.13 Dynamic response plots for viscosity control system.

In a typical case, shown by the solid line, the output viscosity may temporarily increase until the controller has time to correct the fuel heating power. How quickly this occurs may depend on many factors, e.g., the speed of response of the controller and steam supply, or the thermal mass and heat transfer properties in the heat exchanger.

One could also imagine a case where the control system over-compensates slightly and drives the output viscosity below the setpoint, before settling at the desired value. This may give a quicker response away from the high-viscosity output, at the cost of some oscillation around the setpoint before the output settles. A third alternative response could be that the control system is unable to fully compensate for the change in input viscosity; this is shown by the second dotted line. In this case, the output viscosity will remain at some value different from the setpoint; there will be a *steady-state error*. This may be the case, e.g., if the heater is unable to supply enough heat for the new fuel, or if the controller is wrongly tuned.

We will see numerous further examples of system dynamic response later.

1.5 ADVANCED CONTROL ENGINEERING TOPICS

Control engineering is a vast subject area and this text presents only an introduction to control engineering and control systems design. To give the reader an idea of the "bigger picture," this section presents some important topics for potential further study and their relevance.

NONLINEAR CONTROL

All real-world systems have some degree of nonlinearity due to, for example, time delays, hysteresis, or simply that the system has inherent nonlinear characteristics. (We cover some of these aspects in Section 2.4.) Sometimes, the system can be approximated as a linear system, but in many cases it is critical to take the nonlinear characteristics of a system into account in order to achieve acceptable controller performance.

DIGITAL CONTROL

Advances in computer technology in recent years have opened up the way for high-performance digital controllers. There are a number of advantages of using a digital system to apply a control algorithm: (a) Digital controllers are software-based so they can be reprogrammed. (b) They can be interfaced to modern IT systems, giving data robustness and data security. This offers flexibility in terms of changing the controller design, and it also enables the implementation of advanced control strategies, e.g., knowledge-based techniques.

OPTIMAL CONTROL

Optimal control is concerned with optimizing the operation of a system to minimize some cost function. For example, how to get from state A to state

B using the least possible amount of energy. Using mathematical optimization techniques, the controller can be designed to meet such criteria.

ROBUST CONTROL

Robust control is important in systems which are not well defined, such as highly complex plants with internal feedback mechanisms or systems with time-varying properties. A robust controller seeks to employ stability margins in order to maintain system performance for varying conditions.

INTELLIGENT CONTROL

Intelligent controllers are those which use knowledge-based techniques to improve controller performance. This allows the utilization of knowledge of the system or operator experience, and can include using such techniques to tune standard (e.g., PID) controllers, or by employing an intelligent controller. Common techniques used include fuzzy logic and artificial neural networks. One major advantage is that such controllers can be highly nonlinear, giving superior performance in nonlinear plants.

ADAPTIVE CONTROL

As the name suggests, adaptive control techniques seek to adjust the controller performance to the plant behavior. For example, the behavior of ship will depend heavily on the loading, and an adaptive controller will take this into account when determining the controller response. Adaptive controllers can be pretuned, but can also be combined with intelligent control to make the controller self-tuning. The latter can be useful in plant whose properties vary with time, for example, due to wear in the system.

1.6 SOFTWARE FOR CONTROL SYSTEM ANALYSIS AND DESIGN

Software tools are today invaluable in the analysis and design of control systems. Extremely powerful tools exist and these programs are under constant development. The market-leading tool for system analysis and simulation is Matlab/Simulink [1] offered by The Mathworks, Inc. Matlab/Simulink allows both analytic and numerical analysis of dynamic systems and very useful design tools exist in the range of toolboxes (e.g., the control system, robust control, and nonlinear control toolboxes) available for the software.

A free, open-source alternative to Matlab is Scilab [2], developed by a consortium of academic and industrial users. Although Scilab does not contain the same number of sophisticated and specialized features as Matlab, it does have very powerful tools for control engineering. Octave [3] is another open-source software package which also has a range of control-related tools. Its syntax is very similar to Matlab's, however it is predominantly command line-based which many users consider less user-friendly. In laboratory environments, LabView [4] is widely used; this software also includes system simulation capabilities.

In this book, we provide simulation examples using the freely available Scilab software package, along with example code for Matlab. Reproducible examples of the analysis of dynamic systems using this software are included throughout the text. There is plenty of introductory material covering the basic use of Scilab and Matlab on the respective Web sites,[1] hence this will not be covered in detail here.

QUESTIONS

1.1 In the below table, we have listed some examples of control systems and broken them down in terms of the individual system components. Try to think of other automatic control systems you know from your home, your workplace, or elsewhere. In each case, identify:
- the plant and its control objectives;
- the desired output/setpoint;

Plant, Control Objectives, and Setpoint	Control Variable, Actuator	Plant Output and How It Is Measured	External Disturbances, Plant Changes
Ship. Objectives: use ship autopilot to maintain a given heading. Setpoint: desired heading angle (e.g., 35°).	Rudder position, steering gear.	Ship heading measured by compass.	Wind and wave conditions, ship draught, water depth.
Car. Objectives: use cruise control to maintain constant speed. Setpoint: desired speed (e.g., 80 km/h)	Engine fuel flow, carburettor or fuel pump.	Vehicle speed measured by speedometer.	Wind; road conditions; number of people in the car
Diesel engine cooling system. Objectives: use flow control to maintain a given cooling water temperature. Setpoint: desired cooling water temperature (e.g., 60 °C).	Coolant flow, flow control valve.	Cooling water temperature at engine outlet, thermocouple.	Engine load variations, sea water temperature variations.
Diesel generator. Objectives: use fuel control to maintain a fixed engine speed and electrical frequency. Setpoint: desired speed (e.g., 1800 rpm).	Engine fuel flow, fuel pump/ governor.	Engine speed measured by tachometer.	Electric load variations, ambient conditions variations.

---

[1] For Scilab, see the Resources section of the Web site, which contains a number of tutorials, a Wiki page, and references to books. The Mathworks Inc. provides extensive support material on the Documentation Center section of their Web site, including basic use of Matlab as well as the use of the Control Systems Toolbox and Simulink.

- the control variable (the plant input variable that we aim to influence to achieve the desired output);
- the output and how it is measured;
- external disturbances and changes to the process or plant which influence control system performance.

REFERENCES

[1] The MathWorks, Inc. http://www.mathworks.com/products/matlab/.
[2] Scilab Enterprises S.A.S. http://www.scilab.org/.
[3] Octave Project. http://www.octave.org/.
[4] National Instruments Corporation. http://www.ni.com/labview/.

Chapter Two

System Representation in the Time Domain

CHAPTER POINTS

- Marine system types.
- System modeling.
- System realities; nonlinearity.
- First- and second-order systems.
- Standard form of differential equations.
- System identification from test data.
- Example: electric pump drive.

2.1 SYSTEMS AND SYSTEM STUDY

A system is a collection of interconnected components in which there is a specified set of dynamic variables called inputs (or excitations) and a dependent set called outputs (or responses). System analysis involves the subdivision of a system into its constituent parts or elements in an effort to either clarify or to enhance the understanding of the system and its characteristics.

In the context of studying the dynamics and control of a process or plant, a model means a representation of the plant behavior in terms of mathematical statements, usually as differential equations, and the arrangement of them in a convenient form. The representation can be based on:

- observed behavior;
- natural physical laws; or
- a combination of the two.

2.2 MARINE SYSTEM TYPES

There are several types of systems commonly used in marine applications. The most common are mechanical, hydraulic, pneumatic, and electrical systems. *Mechanical systems* and their applications include, for example, linkage mechanisms, gearing systems, and spring and damper systems. Very often systems have some mechanical aspects, even if predominatly of a different type, for example, fluid or electrical systems. One example is the steering gear system illustrated in Figure 2.1.

Marine Systems Identification, Modeling, and Control. http://dx.doi.org/10.1016/B978-0-08-099996-8.00002-1

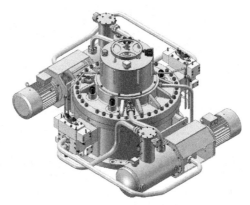

FIGURE 2.1 Porsgrunn steering gear™: a combined electrical, mechanical, and hydraulic system. (Figure courtesy of MacGregor Porsgrunn AS, reprinted with permission.)

Hydraulic systems are extensively used in marine control. Their advantages include:
- Very large forces can be obtained from high-pressure hydraulic systems.
- Hydraulic components are compact with high power densities.
- Hydraulic systems can withstand shock or vibrations and work in dirty or contaminated environments.
- Hydraulic components can be very fast-reacting.
  There are three major elements in hydraulic systems:
- Pumps that supply the high-pressure liquid for the system.
- Valves that control the direction and rate of flow.
- Actuators that utilize the high-pressure fluid to accomplish the required rotation or translation.

Electrical systems are widely used for control purposes and for energy distribution. Electrical systems are versatile and convenient, with signals easily transmitted through electric cables. Such systems provide power generation, transmission, and utilization with very low losses (high component efficiencies). Electronic control systems benefit from very fast system response times.

Pneumatic systems are distinguished from hydraulic systems in that the working medium is compressible, typically air. The advantages of using air as the working fluid include:
- Air is readily available and after use it can be exhausted to the atmosphere.
- Air is not flammable and nonpolluting, hence there are no risks in the case of leakage.
- The characteristics of air are not heavily influenced by temperature.

2.3 SYSTEM MODELING

The following sections show how we can derive mathematical models that can represent dynamic systems of different types. In all cases of modeling, a series of assumptions need to be made, in order that a mathematical representation be possible and the model not be overly complex. For example, for an oil tanker the resistance (or friction) due to air may be orders of magnitude lower than that of the water. We may therefore be able to ignore air resistance without introducing too much error in the model. For a planing speed boat, however, this may not be the case.

As we look at the modeling of different types of systems, we will also examine the modeling assumptions necessary to enable the model derivation.

2.3.1 Linear time-invariant models

We will focus on linear models here, as these are mathematically convenient to work with and to show the principles of system modeling and control. In practice, all real systems are nonlinear to a higher or lesser degree, but often a linear representation is sufficiently accurate to do a useful analysis. (However see the notes on nonlinearity in Section 2.4.)

A physical system is said to be linear time invariant (LTI) when the equations governing it are linear differential equations having constant coefficients. The requirements for a model to be LTI are:

- Additivity: if $f(x_1) + f(x_2) = f(x_1 + x_2)$, then a function fulfills the requirement of additivity.
- Homogeneity: if $k \cdot f(x) = f(kx)$, then a function fulfills the requirement of homogeneity.
- Time invariance: if a time-shifted input produces a time-shifted (but otherwise identical) output, the function is time invariant.

The features of additivity and homogeneity allow the use of superposition and scaling in the system analysis.

2.3.2 Initial conditions and inputs

The initial state of the system variables represents the initial conditions, taken at the instant of time when the dynamic problem starts. These must be known in order to solve the differential equation. Such initial states can be, for example, the position of a mass, the level of fluid in a tank, or the voltage across a capacitor at the start of the simulation. Conveniently, because all the models we will be looking at here are LTI (as described above), we can often simply set the initial conditions to zero.

INPUT TYPES

Inputs to a system may in theory have any form; however, we will concentrate on four common input types: impulse, step, ramp, and sinusoidal inputs.

An *impulse input* is a very high pulse applied to a system over a very short time (i.e., it is not maintained). That is, the magnitude of the input approaches infinity while the time approaches zero. An example of an impulse input is a hammer striking a bell; the bell will experience a very high force but over a very short time. The impact will leave the bell swinging back and forth, hence the impulse input has delivered a finite amount of energy to the system. Another example is a car encountering a bump in the road at high speed; the suspension system will experience an impulse input.

A *step input* is instantaneously applied at some time (typically taken as zero) and thereafter held at a constant level. An example is the voltage applied when closing a switch in an electric circuit, or the fluid flow from a pipe when a valve is opened. Before $t = 0$ the voltage and fluid flow were zero, for $t \geq 0$ they have some constant value.

A *ramp input* increases linearly with time. In theory, the input goes to infinity, however in practice, there is a physical limit or the dynamic problem ends before the input gets too large. An example of a ramp input is the pressure head from a tank which is constantly filling up. If something is connected downstream of the tank, it will experience a ramp change in the pressure.

Figure 2.2 illustrates these three input types. Clearly, in practice neither impulse nor step inputs are physically realizable, since all real systems will have some finite peak magnitude or rise time. However, if the dynamics of the plant are much slower than those of the input, these models will be good approximations. For example, when switching on an electric heater it may take several seconds for it to warm up. In this case, the few milliseconds it takes for the voltage over the input terminals to change are insignificant; for the heater this will appear as a step change.

We will look at the fourth input type, sinusoidal inputs, in Chapter 4.

2.3.3 Modeling fluid systems

When modeling fluid systems, we usually make assumptions as to the linearity of the system. Commonly, this includes: (a) the flow rate through an orifice is proportional to the pressure differential, or head; (b) the fluid is incompressible; and (c) no energy is dissipated in the orifice. This assumes laminar flow, which is not a realistic assumptions in many industrial systems, where turbulence plays a significant role. Many such fluid systems will have nonlinear characteristics. However, for now let us assume that these assumptions hold.

WATER TANK WITH OUTLET
Consider the tank in Figure 2.3, which has water level, h, of 5 m at time $t = 0$ (i.e., $h(0) = 5$). At $t = 0$, the outlet is suddenly opened and the water flows out

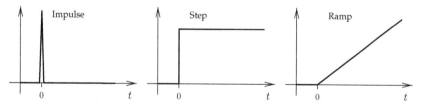

FIGURE 2.2 Types of system inputs.

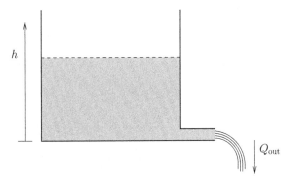

FIGURE 2.3 Water tank with outlet.

of the tank in a rate which is proportional to the instantaneous level of water in the tank.

The water flow out of the tank is therefore $Q_{out} = c \cdot h$, where c is some constant. The rate of change in water volume in the tank can now be expressed as $\dfrac{dV}{dt} = -Q_{out}$. For a constant tank surface area, A, we have $\dfrac{dV}{dt} = A\dfrac{dh}{dt}$. The differential equation for the system is thus

$$\frac{dh}{dt} = -kh,$$

where $k = c/A$. The general solution for this differential equation is

$$h = Ce^{-kt},$$

where C is a constant. (You can confirm this by substituting this into the differential equation.) To find the value of C for a particular system, we need to know the initial conditions. In our case, we know that at $t = 0$ the level of water is $h(0) = 5\,\text{m}$. Substituting for t and h, we get

$$5 = Ce^{-k \cdot 0} \Rightarrow 5 = C,$$

and our solution is therefore

$$h = 5e^{-kt}.$$

Figure 2.4 shows the solution for the water tank level equation we derived. The constant k will depend on the system, in this case, among other things, the size of the tank and of the outlet. A tank with a small outlet will have the same type of response as one with a larger outlet, only slower.

WATER TANK WITH INLET AND OUTLET

Let us look at another hydraulic system, a water tank with an inlet and an outlet as shown in Figure 2.5. We will derive a differential equation describing the output

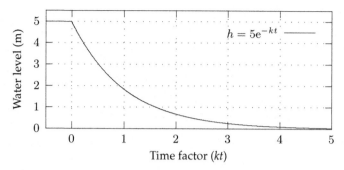

FIGURE 2.4 Solution of the derived equation for water tank level.

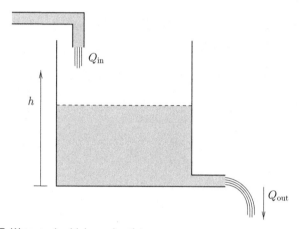

FIGURE 2.5 Water tank with in- and outlet.

flow from the tank as a function of the input flow and the tank design parameters. The flow rate at the outlet is, as above, proportional to the level of water in the tank, i.e.,

$$Q_{out} = kh,$$

where k is a constant. The difference between in- and outflow must equal the rate of change of water volume in the tank, hence

$$A\frac{dh}{dt} = Q_{in} - Q_{out},$$

where A is the area of the water surface. Differentiating the first equation, we have that

$$\frac{dQ_{out}}{dt} = k \cdot \frac{dh}{dt}.$$

Sustituting and rearranging, we get

$$\frac{dQ_{out}}{dt} + \frac{k}{A}Q_{out} = \frac{k}{A}Q_{in}.$$

If we are interested in the water level in the tank rather than the output flow, the same method can be used to derive an expression for h as a function of Q_{in}:

$$\frac{dh}{dt} + \frac{k}{A} \cdot h = \frac{Q_{in}}{A}.$$

We will see how to solve these equations later.

2.3.4 Modeling mechanical systems

Mechanical systems can be modeled using the laws of mechanics. By isolating a part of a system and looking at the resultant forces or moments acting on that part, Newton's laws of motion can be used to describe its dynamics.

There are three main components of mechanical systems: masses, springs, and dampers. For the purpose of modeling, these are commonly assumed to be "ideal" components. An ideal spring is fully reversible (no internal friction), has no mass, and exerts a force proportional to the displacement, or compression, of the spring. An ideal damper has no mass, resists any motion with a force which is proportional to the velocity of that motion, and is irreversible. Other typical assumptions include that there is no external friction (e.g., air resistance), and that external factors such as gravity can be ignored; however, this will depend on the specific case.

One example of a mass-spring-damper system is an engine mount with dampers to reduce vibration. However, numerous physical systems can be modeled in this way, even though one may not instinctively think of them as mass-spring-damper systems. Most structural elements will have some stiffness (the "spring constant") and some element of damping associated with them. For example, a concrete pillar or a support beam can be considered as a very stiff spring.

The principles of linear-acting mechanical systems also apply equivalently to rotating systems, i.e., torsional stiffness and damping.

MASS ON A SPRING
Consider a mass suspended from an ideal spring, as illustrated in Figure 2.6. At equilibrium, we define the displacement $y = 0$ and there is by definition no resultant force on the mass. For any additional extension or compression, the spring reacts with a force, F_s, in the opposite direction:

$$F_s = -ky,$$

where k is the spring constant.

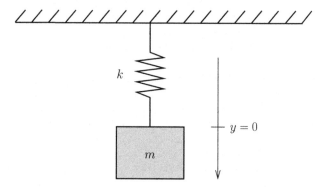

FIGURE 2.6 Mass-spring system.

But we also know that any resultant force, F_{res}, on the mass causes an acceleration defined by $F_{res} = ma$, where m is the mass and a is the acceleration of that mass, as defined by Newton's second law. Assuming that no other forces act on the mass, the resultant force is equivalent to F_s. Since acceleration is the second derivative of the displacement, we have:

$$m\frac{d^2y}{dt^2} = -ky.$$

This is the differential equation for the system and can also be written in the form:

$$\frac{d^2y}{dt^2} + \frac{k}{m}y = 0.$$

This is a second-order linear differential equation with constant coefficients.

TIME-DOMAIN SOLUTION FOR THE MASS-SPRING SYSTEM

We will look at techniques for solving differential equations later. However, let us see what type of response we can expect for the mass-spring system.

For example, assume that we have a mass-spring system as above with a mass of 2 kg and a spring constant of 79 kg/s² and we pull the mass downward 5 cm and then release it. How will it respond? Our system equation with $k = 79$ and $m = 2$ is then

$$\frac{d^2y}{dt^2} + \frac{79}{2}y = 0.$$

With the given initial conditions ($y = 0.05$ m at $t = 0$), the solution for this differential equation is

$$y = 0.05 \cos 6.28t.$$

So the system oscillates with a frequency of 6.28 rad/s ($= 1$ Hz) and amplitude 0.05 m, as shown in Figure 2.7. Note that the frequency of oscillation will not

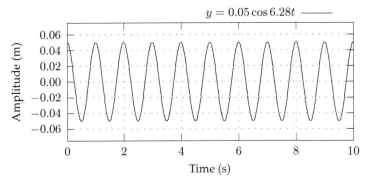

FIGURE 2.7 Time response of mass-spring system.

depend on how far down we pull the mass, only on the mass and the spring constant. Note also that the amplitude is constant, but in real life we would expect the amplitude to decrease over time. This is because our model does not take into account the losses due to frictional resistance in the spring and air resistance acting on the mass as it moves.

MASS ON SPRING AND DAMPER

Developing the previous model further with the addition of a damper element, we get a mass-spring-damper system as shown in Figure 2.8. The spring has the same properties as before, $F_s = -ky$. The damper resists any motion with a force proportional to the velocity of that motion, hence the damper force $F_d = -C\dfrac{dy}{dt}$, where C is the damper constant. In addition, we can assume that there may be another force acting on our system: an external input force, F_i.

Again we use Newton's second law and express the resultant force on the mass as mass times acceleration:

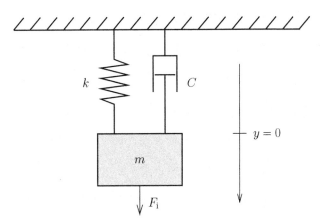

FIGURE 2.8 Mass-spring-damper system.

$$m\frac{d^2y}{dt^2} = -C\frac{dy}{dt} - ky + F_i.$$

This can also be written as

$$\frac{d^2y}{dt^2} + \frac{C}{m}\frac{dy}{dt} + \frac{k}{m}y = \frac{F_i}{m}.$$

We can see that the left-hand side (LHS) of the differential equation is associated with the "system" and the right-hand side (RHS) with the "input." The effect of the additional damping component compared with the mass-spring system above is the additional differential term that is now present in the system equation.

For an example of how the mass-spring-damper system responds to an external excitation, see the plot on page 31. We will come back to mass-spring-damper systems later.

2.3.5 Modeling electrical systems

We can model electrical systems based on Kirchhoff's laws, which can have two forms: (a) the sum of all the potential (i.e., voltage) differences around a closed circuit is zero or (b) the sum of the currents flowing into a junction equals the sum of the currents flowing out of that junction.

Electric circuits consist of three main elements: resistors, coils, and capacitors. As with mechanical systems, we assume that these components and the electric circuit are "perfect," that is: no energy is dissipated in the coil or the capacitor, the resistors are purely resistive (i.e., have no capacitive or inductive parts), and there is no resistance in the circuit connectors (cables, etc.).

LC CIRCUITS

Let us consider a simple LC circuit, as shown in Figure 2.9. The voltages over a coil, V_L, and a capacitor, V_C, can be written as

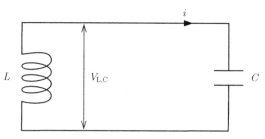

FIGURE 2.9 LC electric circuit.

$$V_L = L\frac{di}{dt},$$

$$V_C = \frac{q}{C},$$

where i is the electric current and q is the capacitor charge. Since $V_L = V_C$ in this circuit, we can write

$$L\frac{di}{dt} - \frac{q}{C} = 0.$$

Knowing that electric current is the rate of change in charge, $i = \frac{dq}{dt}$, we get a differential equation representing this system:

$$\frac{d^2q}{dt^2} - \frac{q}{LC} = 0.$$

LCR CIRCUITS

In this case, we have an LCR circuit as shown in Figure 2.10, which comprises a resistor, a coil, and a capacitor. In addition, there is an input voltage, V_{in}, exciting the system. As above, the voltage over each element in the circuit can be expressed as follows:

$$V_R = Ri = R\frac{dq}{dt},$$

$$V_L = L\frac{di}{dt} = L\frac{d^2q}{dt^2},$$

$$V_C = \frac{q}{C}.$$

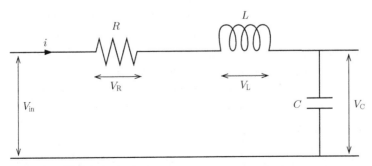

FIGURE 2.10 LCR electric circuit.

In this circuit, $V_{in} = V_R + V_L + V_C$, and we get:

$$V_{in} = L\frac{d^2q}{dt^2} + R\frac{dq}{dt} + \frac{q}{C}.$$

2.3.6 Other system types

The above sections show how we can model some common engineering systems. However, there are numerous other systems which can be modeled in a similar way. Basically, whenever there is a rate of change (the derivative) of a function involved, one ends up with a differential equation.

Examples of other common system types include:

- thermal systems, in which heat transfer depends on the system temperatures;
- chemical or process engineering systems, in which a reaction rate depends on the presence of particular species;
- biological systems, in which the growth rate of, e.g., bacteria depends on the existing population; and
- radioactive decay, in which a substance decomposes at a rate proportional to the amount present.

All these systems can be modeled using an approach similar to that used above. In practice we will often encounter combined systems, e.g., mechanical systems with elements of fluid flow (such as hydraulic systems), or combined mechanical-electrical systems (such as electric machines).

EXAMPLE: SHIP PROPULSION

A ship is accelerated by the thrust force from the propeller, F_t, while the flow resistance, F_r, depends on the instantaneous vessel speed, v. We can assume that the flow resistance is proportional to speed, $F_r(t) = k_1 \cdot v(t)$. The thrust force from the propeller may be assumed proportional to its rotational speed, n, such that $F_t(t) = k_2 \cdot n(t)$.

We can now set up the relationship between the rotational speed of the propeller and the ship speed using Newton's laws of motion ($\sum F = m \cdot a$):

$$F_t - F_r = m \cdot \frac{dv}{dt},$$

$$k_2 n(t) = m \cdot \frac{dv}{dt} + k_1 v(t).$$

2.4 SYSTEM REALITIES

We have been looking at linear system models and we need to remember that they are just that: *models*. The equations we set up for a plant and a controller are approximations to reality and may often contain some nonlinearities, or they may only be linear within a certain range.

Nonlinearity can be continuous or discontinuous. We may be able to linearize continuous nonlinearity but we cannot linearize discontinuous nonlinearity.

2.4.1 Continuous nonlinearity

Figure 2.11a shows the force from a spring for varying displacement. Although the function is not exactly linear we have made a linear approximation of $y = kx$.

In many cases it is not easy to make the appropriate assumptions to achieve linearity; however, it is possible to write down the nonlinear equations and then linearize about what is known as an operating point. For a nonlinear relationship

$$y(t) = f(x(t)),$$

a good approximation, provided that deviation from the operating point is small, is

$$y = f(x_0) + \frac{df}{dx}\bigg|_{x=x_0} (x - x_0) = y_0 + M(x - x_0),$$

where

$$M = \frac{df}{dx}\bigg|_{x=x_0}.$$

This is illustrated in Figure 2.11b. Depending on the system to be modeled, this may give an acceptable accuracy for a given operating range, e.g., between the values x_a and x_b.

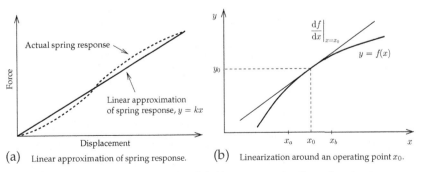

(a) Linear approximation of spring response.

(b) Linearization around an operating point x_0.

FIGURE 2.11 Continuous nonlinearity. (a) Linear approximation of spring response. (b) Linearization around an operating point x_0.

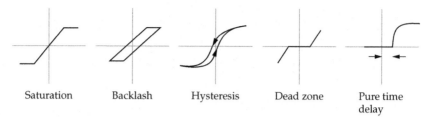

FIGURE 2.12 Discontinuous nonlinearity.

2.4.2 Discontinuous nonlinearity

Some discontinuous nonlinearities are shown in Figure 2.12. We cannot linearize these.

Saturation is a relevant practical aspect of a wide range of systems. Suppose that we have a bow thruster as part of a position control system on a ship. The bow thruster has a maximum power output and is therefore subject to saturation: whatever the demand, the actuator is limited by this maximum power output value. Hence, the response is discontinuous and we cannot linearize for this effect. However, we can ignore it if we say that the operation of our system will be within the linear region.

Backlash and *deadzone* are common on systems with gears or mechanical linkages.

Hysteresis is a property in which the input/output characteristics are different depending on whether the input is increasing or decreasing. This can be considered as a kind of memory effect.

Pure time delay is found in, e.g., fluid systems with long pipes. It is characterized by an output which responds to an input in a regular manner, but after a length of time.

2.5 STANDARD FORM OF DIFFERENTIAL EQUATIONS

As we have seen, many physical systems can be modeled by differential equations which fit into a standard or general form:

- First order:

$$\tau \frac{dy}{dt} + y = Ku(t).$$

- Second order:

$$\frac{d^2 y}{dt^2} + 2\zeta\omega_n \frac{dy}{dt} + \omega_n^2 y = K\omega_n^2 u(t).$$

Having the differential equation in this standard form, we can get some information about the system in question. The right-hand side of the equation is associated with the input, and the left-hand side is the system equation. The input to the system is a function $u(t)$, and the output is y. Setting the right-hand side to zero and solving the differential equation for y gives the "free response" of the system (the response of the system without a "forcing function" or input). This is also known as the homogeneous form of the differential equation.

The coefficients appearing in the equations and their physical importance are as follows:

- K is a constant scaling factor, the system gain. The gain of a system is the ratio of the steady-state system output to the size of the input.
- τ is the time constant of a first-order system. It has a time unit (e.g., seconds) and determines how quickly the system responds to an input. For example, in the first water tank system above, a large tank with a small outlet would have a large τ (it will react slowly), whereas a smaller tank will have a smaller τ and empty more quickly.
- ω_n is the undamped natural frequency of a second-order system. This is the frequency (in rad/s) at which the system will oscillate without a damping term and without a forcing function (input). We saw this in the mass on a spring example above.
- ζ is the damping ratio of a second-order system. The damping ratio is dimensionless and determines the rate at which oscillations die away. It can have a value from zero to infinity, and the special case of $\zeta = 0$ was in fact the case in the mass on a spring example; in that case, oscillations will continue forever.

We will see more of these variables and their relevance later.

2.5.1 Time response of first-order systems

First-order systems are convenient in that the shape of the system response to a given input will be similar; what will change is the final value (depending on the gain K) and the speed of response (depending on the time constant τ).

Figure 2.13 shows the response of some first-order systems with varying values of K and τ for a unit step input. The upper three graphs show systems with a fixed time constant and varying gain; the lower graphs show responses for a fixed gain and varying time constant.

2.5.2 Time response of undamped second-order systems

For the mass on a spring example above (page 21), we first modeled the "free response" (i.e., without any sustained input). The system equation is

$$\frac{d^2y}{dt^2} + \frac{k}{m}y = 0.$$

We can compare this with the standard form:

$$\frac{d^2y}{dt^2} + 2\zeta\omega_n\frac{dy}{dt} + \omega_n^2 y = K\omega_n^2 u(t).$$

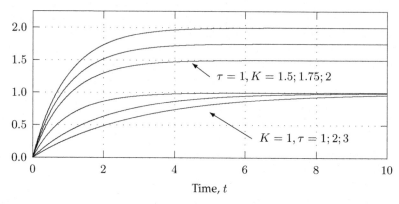

FIGURE 2.13 First-order system response to a step input with varying coefficients.

Comparing the y terms, we see that the natural frequency can be found as $\omega_n^2 = \dfrac{k}{m}$. In this case, there is no damping ($\zeta = 0$) meaning that the system will oscillate at exactly ω_n and the oscillations will not die away with time, since our model did not account for frictional losses and air resistance.

For a mass of 2 kg and a spring constant of 79 kg/s^2, we have

$$\omega_n^2 = \frac{k}{m} = \frac{79}{2} \Rightarrow \omega_n = 6.28\,\text{rad/s}.$$

The natural frequency of this system is 6.28 rad/s (1 Hz) which is the same as we calculated previously.

2.5.3 Time response of damped second-order systems

For the mass-spring-damper example (see page 23), we had:

$$\frac{d^2 y}{dt^2} + \frac{C}{m}\frac{dy}{dt} + \frac{k}{m}y = \frac{F_i(t)}{m}.$$

We can see that, as above, the spring constant k and mass m influence the natural frequency term. The damper constant C influences only the first derivative term, which comprises the damping ratio, ζ, which is what we would expect.

We have

$$\omega_n^2 = \frac{k}{m} \Rightarrow \omega_n = \sqrt{\frac{k}{m}} \quad \text{and}$$

$$2\zeta\omega_n = \frac{C}{m} \Rightarrow \zeta = \frac{C}{2m\sqrt{\frac{k}{m}}} = \frac{C}{2\sqrt{km}}.$$

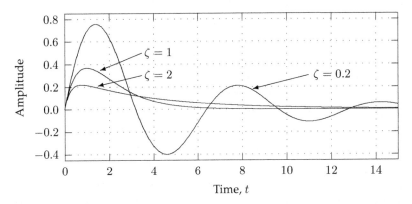

FIGURE 2.14 Second-order system response to an impulse input with varying damping ratio.

Solving the second-order equation for the mass-spring-damper system in the time domain is not straightforward, so we will leave the details of that for later. It is however worth noting that the type of response of a second-order system will depend on the damping ratio, ζ. This is different to first-order systems, which will always have the same type of response, only with different speed of response and gain.

Figure 2.14 shows the response of a second-order system to an impulse input (e.g., if we were to strike the mass in a spring-mass-damper system with a hammer) for varying damping ratio values.[1] For a high damping ratio ($\zeta = 2$), the strike is effectively damped with only a small amplitude which slowly returns to the original position. For a low damping ratio ($\zeta = 0.2$), the mass oscillates for a while before returning to its original position. The special case of $\zeta = 1$ is called critical damping; in this case the mass returns to its original position as quickly as possible without oscillating.

If $\zeta = 0$, we have the same case as above (no damping) and the system will continue to oscillate forever at its undamped natural frequency, ω_n. If $0 > \zeta > 1$, the system will oscillate before it settles, but at its *damped natural frequency*, ω_d. The relationship between undamped and damped natural frequencies is

$$\omega_d = \omega_n \sqrt{1 - \zeta^2}.$$

2.6 SYSTEM IDENTIFICATION FROM TEST DATA

In some cases, the system we are dealing with is too complex to model accurately, or we simply don't know enough about it to be able to model it accurately from first principles. For example, if you are designing a cruise control system for

---

[1] One can envisage the damping ratio as equivalent to the thickness of the oil in a dashpot viscous damper, as illustrated in Figure 2.8. A thick oil will give more damping and a higher damping ratio. A thin oil produces less damping and a lower ζ value.

a car, it would be very difficult and laborious to develop accurate submodels for the combustion engine, transmission system, and the vehicle dynamics. In such a case, one may use experimental test data to arrive at an overall system model, by imposing, e.g., a step change in the gas pedal position and measuring the speed change of the vehicle. Provided that we have access to the system to perform such tests and can measure the data with sufficient accuracy, this allows us to create a "black box" model of the system without knowing the details of its inner workings.

We saw the characteristics of first-order system responses in the examples above. In response to a step change in the input, a first-order system will respond in an exponential manner, eventually settling at a new steady-state value. Figure 2.15 illustrates this situation.

The coefficients required to fit this data to a first-order model in the standard form are the steady-state gain K and the time constant τ. The gain is equal to the steady-state change in the process variable divided by the change in the input:

$$K = \frac{\Delta y}{\Delta u}.$$

In this example, the gain is equal to one. Note that the gain can have a negative value, and it also has a unit depending on the system (e.g., K/mA for a temperature control system or Hz/% for an engine speed controller).

The time constant, τ, is equal to the time it takes for the output variable to reach 63% of its final, steady-state value. Hence,

$$\tau = t_{63\%} - t_{\text{start}},$$

where t_{start} is the time when the step change is imposed and $t_{63\%}$ is the time at which y has reached a value of $y_{\text{start}} + 0.63 \cdot \Delta y$. The time constant is always positive and has a time unit (seconds, minutes, hours, ...).

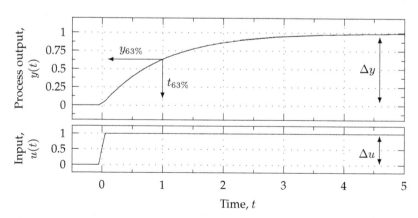

FIGURE 2.15 First-order system identification from test data.

In this way, it is possible to obtain a mathematical representation of a system, or one component within a larger system, even if it cannot be easily modeled. It is possible to fit test data to higher-order models, but it is considerably more complex and will not be covered here.

2.7 EXAMPLE: MODELING AN ELECTRIC PUMP DRIVE

Applying the same mathematical technique as above, we will develop a model of a centrifugal pump drive assumed to be driven by a direct current (d.c.) electric motor, as illustrated in Figure 2.16. In order to model this system, we need mathematical representations in terms of its electrical and mechanical submodels.

The electric motor consists of a stator (the stationary part) with field coils (or permanent magnets) maintaining a constant magnetic field around the armature. The armature contains windings (the armature coils) allowing the magnetic field from the armature to be varied, through varying the current through the windings. This is done by controlling the voltage over the motor terminals.[2]

ELECTRICAL MODEL

From an electrical point of view, a motor consists of a winding which has an inductance L and a small resistance R. This is illustrated in Figure 2.17. These generate a voltage related to the current, i, such that:

$$V_R = iR,$$

$$V_L = L\frac{di}{dt}.$$

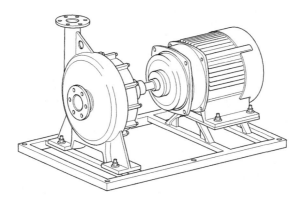

FIGURE 2.16 Centrifugal pump with electric drive.

---

[2] This method of controlling the d.c. motor is known as armature control. It is also common to use field control, in which a constant magnetic field is maintained in the armature and the strength of the field coil is varied.

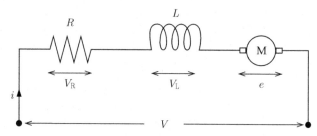

FIGURE 2.17 Electric circuit of a d.c. motor.

When the motor is turning there is also a back voltage, the electromotive force (emf), e, produced in the coil. The emf is proportional to the angular rotor velocity:

$$e = k_e \frac{d\theta}{dt},$$

where k_e is the motor constant. Therefore, if we apply a voltage, V, across the motor terminals we get:

$$V = V_R + V_L + e = iR + L\frac{di}{dt} + k_e\frac{d\theta}{dt}.$$

MECHANICAL MODEL

From a mechanical point of view, the motor consists of a driving (electromagnetic) torque, T_e, which drives a rotor assembly (including the impeller) with an inertia of J. The electromagnetic torque is proportional to the armature current by the armature constant, k_a:

$$T_e = k_a \cdot i.$$

The load acts on the impeller assembly with a load torque T_L. Frictional forces also oppose the rotation of the rotor assembly, and we assume that the torque due to friction is proportional to the motor angular velocity:

$$T_f = k_f \frac{d\theta}{dt},$$

where k_f is a friction factor. We assume that these are the only torques acting on the rotor.

Equivalent to Newton's second law for linear-acting systems, the resultant torque (i.e., the sum of all torques) acting on the rotor assembly will determine its angular acceleration. The equivalent of mass in linear systems is inertia for rotating systems, and we get

$$J\frac{d^2\theta}{dt^2} = \sum_n T_n,$$

$$J\frac{d^2\theta}{dt^2} = T_e - T_L - T_f,$$

$$J\frac{d^2\theta}{dt^2} + k_f\frac{d\theta}{dt} = T_e - T_L.$$

SUMMARY

To summarize, we have two equations which model the electric pump drive:

$$iR + L\frac{di}{dt} = V - k_e\frac{d\theta}{dt},$$

$$J\frac{d^2\theta}{dt^2} + k_f\frac{d\theta}{dt} = T_e - T_L, \quad \text{where}$$

$$T_e = k_a \cdot i.$$

The inputs to the system are the voltage applied to the motor terminals, V, and the load torque T_L. We will use these later to develop control systems for the motor.

QUESTIONS

2.1 Consider a block of metal at initial temperature T_m in an atmosphere with a constant ambient temperature T_a. By Newton's law of cooling, the rate of heat loss from the block is proportional to the temperature difference. Write down a differential equation to model this system. (Use k for the proportional constant.) State any assumptions you make.

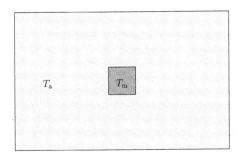

If the ambient temperature, T_a, is $0\,°\text{C}$ and the initial temperature of the block is $50\,°\text{C}$, solve the differential equation and calculate the temperature of the block after $t = \dfrac{1}{k}$ s.

2.2 Consider the capacitor-resistor circuit shown below. A capacitor is charged to 5 V. At time $t = 0$, the capacitor is switched to discharge through the resistor. Using the equations, relate current, i, to voltage, v, for the capacitor and resistor.

$$i(t) = C\frac{dv}{dt} \quad \text{(capacitor)}$$

$$i(t) = \frac{v(t)}{R} \quad \text{(resistor)}$$

Confirm that the voltage after time $t = CR$ is 1.84 V.

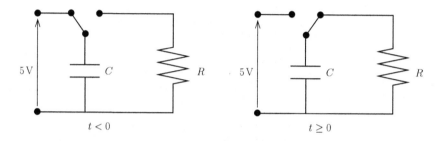

$t < 0$ $t \geq 0$

2.3 In a mass-spring-damper system, if the mass is 1 kg, the damper constant is 3 kg/s, and the spring constant is 2 kg/s², what is the natural frequency and the damping ratio? If we were to pull the mass to a certain point and release it, how would you expect the system to respond?

Chapter Three

System Transfer Functions

CHAPTER POINTS

- Laplace transforms and their properties.
- First- and second-order systems in the Laplace domain.
- Poles and zeros on the s-plane.
- Transient response of first- and second-order systems.
- Electric pump drive in the s-domain.

We have seen that physical systems can be modeled by first- or second-order differential equations. One of the problems with differential equations is that they are difficult to manipulate mathematically. By using Laplace transforms, we can greatly simplify the maths; instead of dealing with integration we can manipulate equations algebraically. This chapter introduces Laplace transforms, then goes on to look at the dynamic properties of various systems, using their transfer functions, i.e., their representation in the Laplace (s) domain.

3.1 LAPLACE TRANSFORMS

The Laplace transform, $F(s)$, of a function $f(t)$ defined for $t > 0$ is given by

$$F(s) = \int_0^\infty e^{-st} f(t) \, dt,$$

where s is a complex number, $s = \sigma + j\omega$, and the inverse Laplace transform is given by

$$f(t) = \int_{\sigma - j\omega}^{\sigma + j\omega} F(s) e^{st} \, ds.$$

Marine Systems Identification, Modeling, and Control. http://dx.doi.org/10.1016/B978-0-08-099996-8.00003-3

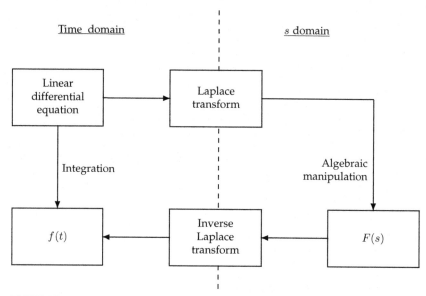

FIGURE 3.1 Solving differential equations in the time and Laplace domains.

The box lists the mathematical definition of the Laplace and inverse Laplace transforms, however we won't worry too much about the details of this at the moment. In practice we will usually look up Laplace transforms from a table rather than calculate them. (A table of common Laplace transforms and inverse transforms is given in Appendix A.)

Figure 3.1 illustrates the usefulness of the Laplace transform. We know that a linear differential equation can be integrated to get the solution in the time domain. However, if the differential equation is complex, it is in many cases easier to take the Laplace transform, manipulate the (transformed) equation algebraically, and then take the inverse Laplace transform to get the time-domain solution.

3.1.1 Properties of Laplace transforms

The list below presents key properties for the Laplace transform which will help us manipulate them; we will refer back to this later when we look at some examples. Note the notation used: lower-case function names are commonly used for the time-domain function and upper-case names when transformed into the Laplace domain. Hence, the Laplace transform of $f(t)$ becomes $F(s)$, and $\mathcal{L}[y(t)] = Y(s)$.

For Laplace transforms, the following properties apply:

1. Linearity:

$$\mathcal{L}[f_1(t) + f_2(t) + \cdots] = \mathcal{L}[f_1(t)] + \mathcal{L}[f_2(t)] + \cdots$$

2. Linearity:

$$\mathcal{L}\left[a \cdot f(t)\right] = a \cdot \mathcal{L}\left[f(t)\right]$$

3. Differentiation/integration with respect to time:

$$\mathcal{L}\left[\frac{df}{dt}\right] = sF(s) - f(0)$$

$$\mathcal{L}\left[\frac{d^2 f}{dt^2}\right] = s^2 F(s) - sf(0) - \frac{df}{dt}\bigg|_{t=0}$$

$$\mathcal{L}\left[\int f(t)\,dt\right] = \frac{1}{s}F(s)$$

4. Initial value theorem:

$$\lim_{t \to 0} f(t) = \lim_{s \to \infty} sY(s)$$

5. Final value theorem:

$$\lim_{t \to \infty} f(t) = \lim_{s \to 0} sY(s)$$

Notice in particular point 3, which shows that the operations of differentiation and integration in the time domain is reduced to multiplying or dividing by s in the Laplace domain. This makes it much easier to manipulate and study the system equations. We will also come back to the initial value theorem and the final value theorem later.

3.1.2 Laplace transform example

The best way to illustrate the use of Laplace transform is through an example. Consider the mass-spring-damper system we looked at in Section 2.3.4 (page 21). The differential equation for the system is

$$m\frac{d^2 y}{dt^2} + C\frac{dy}{dt} + ky = u(t).$$

Taking Laplace transforms, we get

$$\mathcal{L}\left\{m\frac{d^2 y}{dt^2} + C\frac{dy}{dt} + ky\right\} = \mathcal{L}\left\{u(t)\right\},$$

and by properties 1 and 2 from above this is equivalent to

$$m \cdot \mathcal{L}\left\{\frac{d^2 y}{dt^2}\right\} + C \cdot \mathcal{L}\left\{\frac{dy}{dt}\right\} + k \cdot \mathcal{L}\left\{y\right\} = \mathcal{L}\left\{u(t)\right\}$$

Applying property 3 gives

$$m \left(s^2 Y(s) - sy(0) - \frac{dy}{dt}\Big|_{t=0} \right) + C \left(sY(s) - y(0) \right) + kY(s) = U(s).$$

We assume that $y(0) = \frac{dy}{dt}\big|_{t=0} = 0$, and get

$$ms^2 Y(s) + CsY(s) + kY(s) = U(s).$$

We want to solve for the output, Y, so we rearrange to

$$Y(s) = \frac{1}{ms^2 + Cs + k} U(s).$$

Here, $U(s)$ is the input. If we use a unit step input, i.e., a force with a magnitude of one suddenly applied at time $t = 0$, this has a transfer function $U(s) = \frac{1}{s}$ (refer to the Laplace transform table in Appendix A). Using values for m, C, and k of 1 kg, 3 Ns/m, and 2 N/m, respectively, we have:

$$Y(s) = \frac{1}{1s^2 + 3s + 2} \left(\frac{1}{s} \right) = \frac{1}{s(s+1)(s+2)}.$$

Remember that $Y(s)$ is the *Laplace transform* (i.e., s-domain solution) of the system output. To get the output in the time domain, we must apply the inverse Laplace transform. In practice, this is best done by manipulating the equation so that it contains the terms in the same format as they appear in the Laplace transform table (if this is possible). In the above equation, we will separate the terms of the denominator by using partial fractions. We want the equation in the form of the right-hand side of the identity

$$Y(s) = \frac{1}{s(s+1)(s+2)} = \frac{a}{s} + \frac{b}{(s+1)} + \frac{c}{(s+2)},$$

and for that we need to calculate the values of a, b, and c. Multiplying through by the denominator of the left-hand side gives

$$1 = a(s+1)(s+2) + bs(s+2) + cs(s+1),$$
$$1 = (a+b+c)s^2 + (3a+2b+c)s + 2a.$$

The left-hand side of this equation reads $0s^2 + 0s + 1$. We can therefore equate coefficients for the different powers of s. For the constant terms, we have

$$1 = 2a \quad \Rightarrow \quad a = \frac{1}{2}.$$

For the s^1 terms:

$$0 = 3a + 2b + c \quad \Rightarrow \quad c = \frac{-3}{2} - 2b.$$

For s^2:

$$0 = a + b + c \quad \Rightarrow \quad c = \frac{-1}{2} - b.$$

Equating expressions for c, we get

$$\frac{3}{2} + 2b = \frac{1}{2} + b \quad \Rightarrow \quad b = -1$$

$$c = \frac{1}{2}.$$

Therefore, the original equation can be written as

$$Y(s) = \frac{1}{2s} - \frac{1}{(s+1)} + \frac{1}{2(s+2)}.$$

The above technique is called "partial fractions." This is described in more detail in Appendix B.2 and further examples are given.

Looking at the Laplace transform table we can see that the inverse transform of $\frac{1}{s}$ is the unit step, i.e., 1 in the time domain, and that the inverse transform of $\frac{1}{s+\alpha}$ is $e^{-\alpha t}$. Remembering property 2 of Laplace transforms, we can take the multiplying factors (i.e., $\frac{1}{2}$, -1, $\frac{1}{2}$ in this case) outside of the transformations, to get:

$$y(t) = \frac{1}{2} - e^{-t} + \frac{1}{2}e^{-2t}.$$

The graph of $y(t)$ is shown in Figure 3.2. This graph can be produced in Scilab or Matlab with the following code:

```
t = 0:0.01:6;
y = 1/2 - exp(-t) + 1/2*exp(-2*t);
plot(t,y);
```

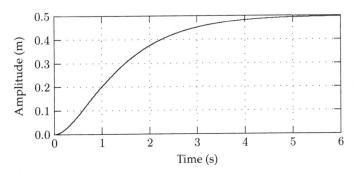

FIGURE 3.2 Time response for mass-spring-damper system.

3.2 TRANSFER FUNCTIONS IN THE s-DOMAIN

It was shown in Chapter 2 that the standard form for a first-order differential equation is

$$\tau \frac{dy}{dt} + y = Ku(t).$$

The Laplace transform of this equation, assuming that $y(0) = 0$, is

$$s\tau Y(s) + Y(s) = KU(s).$$

This system has a transfer function which relates the output Y to the input U, and we can formulate the transfer function for this system in its standard form.

Transfer function for a first-order system

$$\frac{Y(s)}{U(s)} = \frac{K}{s\tau + 1}$$

For a second-order system, the differential equation in standard form is

$$\frac{d^2 y}{dt^2} + 2\zeta \omega_n \frac{dy}{dt} + \omega_n^2 y = K\omega_n^2 u(t).$$

Working out the Laplace transform of this, assuming that all initial conditions at time $t = 0$ are zero, we get

$$s^2 Y(s) - sy(0) - \left.\frac{dy}{dt}\right|_{t=0} + 2\zeta \omega_n \{sY(s) - y(0)\} + \omega_n^2 Y(s) = K\omega_n^2 U(s)$$

$$Y(s)\left[s^2 + 2\zeta \omega_n s + \omega_n^2\right] = K\omega_n^2 U(s)$$

Rearranging this gives us the transfer function for the second-order system in standard form.

Transfer function for a second-order system

$$\frac{Y(s)}{U(s)} = \frac{K\omega_n^2}{s^2 + 2\zeta \omega_n s + \omega_n^2}$$

3.2.1 System example: First-order system

Consider the simple electrical system shown in Figure 3.3, consisting of a resistor and capacitor. The current flow through a resistor is related to the voltage drop by $i(t) = \dfrac{v_R(t)}{R}$ (Ohm's law) and the current flow through a capacitor is related to the voltage drop across it by $i(t) = C\dfrac{dv_C}{dt}$. The current flowing through both components must be the same (Kirchhoff's law). Assume that initially the capacitor is discharged ($v_C(0) = 0$). We equate the expressions for $i(t)$ to get:

$$\frac{v_R(t)}{R} = C\frac{dv_C}{dt}.$$

Since $v_R(t) = v_{in}(t) - v_{out}(t)$ and $v_C(t) = v_{out}(t)$, the relation between input and output voltage is:

$$v_{in}(t) = RC\frac{dv_{out}}{dt} + v_{out}(t).$$

Taking Laplace transforms:

$$V_{in}(s) = RCsV_{out}(s) - v_{out}(0) + V_{out}(s),$$

$$V_{out}(s) = \left[\frac{1}{sRC + 1}\right]V_{in}(s).$$

Note that the output is related to the input by the term in the square brackets. This term is the transfer function and characterizes the system. If we know the transfer function for any system, we can determine the output by simple multiplication in the Laplace domain. This can be represented in block diagram form as shown in Figure 3.4.

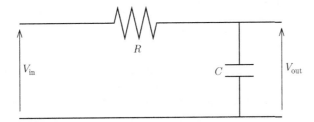

FIGURE 3.3 Electrical system example.

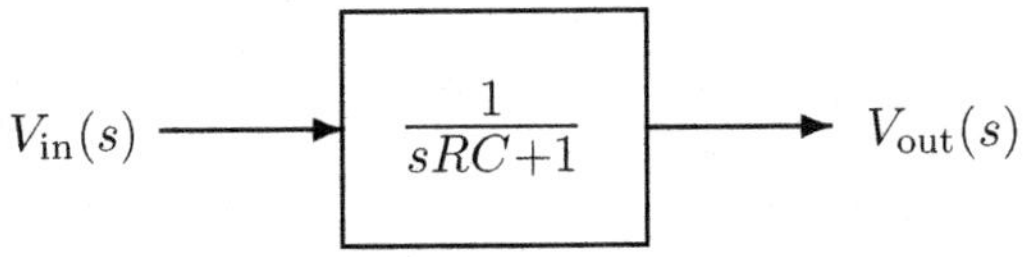

$$V_{in}(s) \longrightarrow \boxed{\frac{1}{sRC+1}} \longrightarrow V_{out}(s)$$

FIGURE 3.4 Transfer function relating input to output signal.

Here, the signal $V_{in}(s)$ is multiplied with the block transfer function to give the output signal $V_{out}(s)$, i.e., it is equivalent to the equation above.

Comparing with the standard form, we see that this is a first-order transfer function, in which the time constant, τ, depends on the values of R and C. Hence, just by looking at the transfer function, we can say something about the dynamics of the system.

Assume a 500 kΩ resistor and a 1 μF capacitor and that the system is excited with 10 V, which is suddenly switched on at $t = 0$. (Remember that initially the capacitor was discharged; if it had some charge the result would be different.) This is a step function, and from the Laplace transform table (page 149) we can see that its transfer function is

$$\mathcal{L}\{\text{step input}\} = \frac{1}{s}.$$

The above is for a *unit* step, so we need to multiply that by 10. (We can do this because the system is linear.) So the equation for V_{out} is:

$$V_{out}(s) = \frac{1}{0.5s + 1} \cdot \frac{10}{s} = \frac{20}{s(s + 2)}.$$

Solving this by partial fractions is fairly straightforward, and we get

$$V_{out}(s) = \frac{10}{s} - \frac{10}{(s + 2)}.$$

Taking inverse Laplace transforms gives the time domain solution:

$$v_{out}(t) = 10 - 10e^{-2t} = 10(1 - e^{-2t}).$$

The time-domain input and response are shown in Figure 3.5, which can be plotted in Scilab and Matlab similarly as above:

```
t = 0:0.01:5;
y = 10 - 10*exp(-2*t);
plot(t,y,[-0.5,0,0,5],[0,0,10,10]);
```

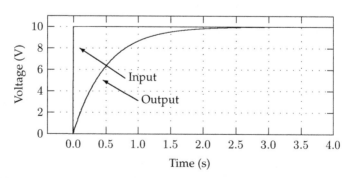

FIGURE 3.5 RC circuit response to a change in input voltage.

However, these software packages conveniently allow us to plot the time domain response directly from the system transfer function and the input in the Laplace domain. For this system, the following Scilab code will produce the same graph:

```
s=%s; // Declare 's' as Laplace operator
sys = syslin('c',1/(0.5*s+1)); // Create system transfer
    function
t = 0:0.01:5; // Create time vector
y = 10*csim('step',t,sys); // Create output data
plot(t,y); // Create plot
```

The equivalent code for Matlab is:

```
s=tf('s'); % Declare 's' as Laplace operator
sys = 1/(0.5*s+1); % Create system transfer function
step(10*sys); % Plot step response for an input of 10
```

3.2.2 System example: Second-order system

We have seen that the second-order differential equation in standard form gives a system transfer function of the form

$$\frac{Y(s)}{U(s)} = \frac{K\omega_n^2}{s^2 + 2\zeta\omega_n s + \omega_n^2}.$$

Referring back to the mass-spring-damper system above (page 40), we had

$$Y(s) = \frac{1}{ms^2 + Cs + k} U(s).$$

This can be rearranged into the standard form:

$$\frac{Y(s)}{U(s)} = \frac{\frac{1}{m}}{s^2 + \frac{C}{m}s + \frac{k}{m}}.$$

Having this, the dynamic properties of a given mass-spring-damper system can be found. The undamped natural frequency ω_n and damping ratio ζ are given by

$$\omega_n^2 = \frac{k}{m} \Rightarrow \omega_n = \sqrt{\frac{k}{m}}$$

$$2\zeta\omega_n = \frac{C}{m} \Rightarrow 2\zeta\sqrt{\frac{k}{m}} = \frac{C}{m} \Rightarrow \zeta = \frac{C}{2\sqrt{km}}.$$

The gain, K, is found by comparing the numerators:

$$K\omega_n^2 = \frac{1}{m} \Rightarrow K = \frac{1}{k}.$$

For example, values of $m = 2\,\text{kg}$, $C = 6\,\text{N s/m}$, $k = 18\,\text{N/m}$, would give:

$$\zeta = \frac{C}{2\sqrt{km}} = \frac{6}{2\sqrt{18 \cdot 2}} = 0.5,$$

$$\omega_n = \sqrt{\frac{k}{m}} = \sqrt{\frac{18}{2}} = 3\,\text{rad/s},$$

$$K = \frac{1}{k} = \frac{1}{18}.$$

3.2.3 Initial and final value theorems

The initial and final value theorems (see page 39) provide a simple method for finding the initial (start) or final value of a dynamic problem. By entering the transfer function for the output (i.e., the system transfer function multiplied by the input), we can get the exact start value and final value directly.

Considering the first-order electrical system form the example above, the output transfer function when excited with a 10 V step input was $V_{\text{out}}(s) = \frac{1}{0.5s + 1} \cdot \frac{10}{s}$. Putting that into the initial value theorem, we get

$$\lim_{t \to 0} f(t) = \lim_{s \to \infty} s \cdot \frac{1}{0.5s + 1} \cdot \frac{10}{s} = \lim_{s \to \infty} \frac{10}{0.5s + 1}.$$

As s goes to infinity, this expression goes to zero. Hence, we know that our output will start at a value of zero. However, we probably knew that already from the initial conditions. More useful is often the final value theorem, which gives us the *final* state of the output:

$$\lim_{t \to \infty} f(t) = \lim_{s \to 0} s \cdot \frac{1}{0.5s + 1} \cdot \frac{10}{s} = \lim_{s \to 0} \frac{10}{0.5s + 1}.$$

As s goes to zero in this expression, the result tends to $\frac{10}{1}$. Hence (if we wait long enough), the output will end up at a final value of 10. This is indeed the same as we can see from the plot on page 44.

Let us try with the spring-mass-damper system from the other example above as well. Using values of $m = 1$, $C = 3\,\text{N s/m}$, and $k = 2\,\text{N /m}$, the transfer function is $\frac{1}{s^2 + 3s + 2}$. If the system is excited with a unit step input, $\frac{1}{s}$, we have

$$\lim_{t \to \infty} f(t) = \lim_{s \to 0} s \cdot Y(s) = \lim_{s \to 0} s \cdot \frac{1}{s^2 + 3s + 2} \cdot \frac{1}{s} = \lim_{s \to 0} \frac{1}{s^2 + 3s + 2} = \frac{1}{2}.$$

We plotted this case on page 41. If we had used an impulse input (Laplace transform $= 1$), we would get:

$$\lim_{t \to \infty} f(t) = \lim_{s \to 0} s \cdot \frac{1}{s^2 + 3s + 2} \cdot 1 = \lim_{s \to 0} \frac{s}{s^2 + 3s + 2} = 0.$$

Hence, when exciting the spring-mass-damper system with an impulse input, it will eventually return to its original state. Therefore, provided that the system in question is stable (we will look at stability later), we can find the final state of the output without solving the system equation. We do not know, however, anything about the dynamics or how much time it will take to get there; the final value theorem only looks at the state at $t = \infty$.

3.3 s-DOMAIN POLES AND ZEROS

We have seen how systems modeled by differential equations can be represented by their transfer function. Let us look in more detail at the properties of the Laplace operator and the significance of a system's representation in the Laplace domain.

Consider a system with a transfer function

$$G(s) = \frac{(s + 1)}{s(s + 2)(s + 3)}.$$

The equation for the transfer function will tend to zero as s tends to -1, so we call $s = -1$ a "zero." Similarly, as s tends to 0, -2 or -3 the equation tends to infinity, and we call these "poles."

The "s" denotes that a function F is a function of the "complex frequency domain" in the same way that t denotes that f is a function of time. Hence, s is a complex number, $s = \sigma + j\omega$, with real part σ and imaginary part ω. We can therefore plot the poles and zeros on an Argand diagram, as in Figure 3.6a. Note that a pole is marked with a cross and a zero with a circle.

The denominator of the transfer function when set to zero constitutes the *characteristic equation* of the system. The solutions of the characteristic equation are the poles of the transfer function. For the system above, all the poles and zeros are real, i.e., they have no imaginary parts. However, we frequently have systems with complex poles. Consider the second-order system

$$G(s) = \frac{(s + 2)}{s^2 + 2s + 3}.$$

To determine the poles of this system, we solve the characteristic equation

$$s^2 + 2s + 3 = 0.$$

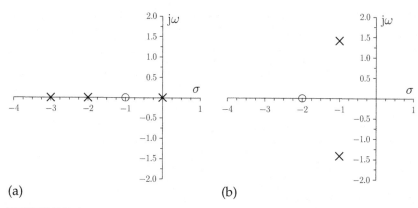

FIGURE 3.6 Argand diagrams showing system poles and zeros (pole-zero maps). (a) Pole-zero map for $G(s) = \frac{(s+1)}{s(s+2)(s+3)}$. (b) Pole-zero map for $G(s) = \frac{(s+2)}{s^2+2s+3}$.

The solutions of this equation are $-1 \pm j1.41$, and they are plotted in Figure 3.6b. Note that the system zero is also different to that above, since the transfer function numerator has changed. Plots such as those shown in Figure 3.6 are known as system *pole-zero maps*.

Pole-zero maps can be plotted easily in Scilab using the plzr command, e.g.:

```
s=%s; // Declare 's' as Laplace operator
sys = syslin('c',(s+2)/(s^2+2*s+3)); // Create transfer function
plzr(sys)
```

For Matlab, use:

```
s=tf('s'); % Declare 's' as Laplace operator
sys = (s+2)/(s^2+2*s+3); % Create transfer function
pzmap(sys)
```

3.3.1 Influence on system dynamics

Linear single-input single-output systems are, in general, modeled by differential equations of the form

$$a_0 \frac{d^n y}{dt^n} + a_1 \frac{d^{n-1} y}{dt^{n-1}} + a_2 \frac{d^{n-2} y}{dt^{n-2}} + \cdots + a_{n-1} \frac{dy}{dt} + a_n y = u(t),$$

where $u(t)$ is the input and $y(t)$ is the output. The characteristic equation of such a system is

$$a_0 s^n + a_1 s^{n-1} + a_2 s^{n-2} + \cdots + a_{n-1} s + a_n = 0,$$

and the complementary function solution is

$$y(t) = c_1 \cdot e^{r_1 t} + c_2 \cdot e^{r_2 t} + \cdots + c_n \cdot e^{r_n t}$$

where $r_1, r_2, \ldots, r_n$ are the roots of the characteristic equation, i.e., the system poles.

We know that any pole, $r = s$, is a complex number, hence for the part of the solution associated with that pole we have

$$e^{rt} = e^{(\sigma + j\omega)t} = e^{\sigma t}e^{j\omega t}.$$

For $t = 0$, we see that $e^{rt} = 1$. If ω is zero, e^{rt} is a real, exponential function, decaying for negative σ, increasing for positive σ, and constant ($= 1$) for $\sigma = 0$. For a positive σ, the output will grow to infinity as time increases. Since $e^{\sigma t}$ is part of the solution, we can draw an important conclusion:

For a system to be stable, i.e., for the output $y(t)$ to arrive at some finite final state as time increases, all the system poles must have negative or zero real parts ($\sigma \leq 0$).

Even if only one of the system poles lies in the positive half plane, that pole will eventually dominate and drive the system to instability.

A further characteristic of physical systems is that complex poles always occur in complex conjugate pairs, as in the example above. If we let σ be zero and take two values of r with equal and opposite imaginary parts, we have $e^{j\omega t}$ and $e^{-j\omega t}$. We know that

$$\cos \omega t = \frac{e^{j\omega t} + e^{-j\omega t}}{2},$$

so the complex conjugate pair represents an oscillatory signal with increasing frequency as ω increases. For any value of r with nonzero real and imaginary parts, the signal will therefore be an exponentially damped or exponentially increasing sinusoid.

Figure 3.7 shows an Argand diagram with some possible pole placements and the contribution from that pole on the system response. Starting with the poles on the imaginary axis, the pole at the origin ($s = 0$) is simply a constant with no dynamics associated with it. The other pole has zero real part, hence no exponential decay or increase, and nonzero imaginary part, giving a pure sinusoidal (harmonic) output. The two poles in the right-hand plane both give an unstable system response, one with an oscillatory response (due to nonzero imaginary part) and one with a purely exponential output.

The poles in the left-hand plane give stable system responses, i.e., the system will reach a finite, steady solution. As one moves toward the left (decreasing σ), the speed of response increases due to the $e^{\sigma t}$ term. The value of $j\omega$ influences the frequency of oscillation; a higher value gives higher oscillation frequency.

Pole placement is of high importance in system study, and we will come back to this in more detail later. Adding a control circuit to a system influences the pole placement, and therefore the dynamic behavior of the system. By using a controller, one may even make a stable system out of an unstable one, by

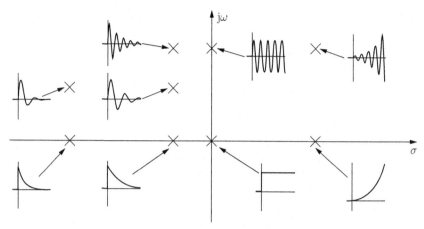

FIGURE 3.7 Influence of pole placement on system dynamic response. (Second poles in conjugate pairs not shown.)

"moving" the poles from the right-hand plane into the left-hand plane. (However a poorly designed controller may also do the opposite.)

3.4 TRANSIENT RESPONSE OF FIRST-ORDER SYSTEMS

Now that we are comfortable with transfer functions and system models, let us look at the dynamic characteristics of some common system types. A first-order system is characterized by the differential equation

$$\tau \frac{dy}{dt} + y = Ku,$$

where u is the input and y is the output. (Remember that τ is the time constant for the system.) The transfer function is

$$\frac{Y(s)}{U(s)} = \frac{K}{(\tau s + 1)}.$$

We will look at the transient response based on three different inputs: the unit impulse (also known as the Dirac delta function), the unit step, and the unit ramp. (The characteristics of these were covered in Section 2.3.2.)

3.4.1 Impulse response

The unit impulse is defined as a pulse at $t = 0$ with infinitesimally small duration. The unit impulse is known (the area under the impulse input graph is unity) so that we can make quantitative analysis. The Laplace transform of the unit impulse is 1. Substituting $U(s) = 1$ in the equation above gives the output transfer function

$$Y(s) = \frac{K}{(\tau s + 1)} = \frac{K}{\tau} \frac{1}{(s + 1/\tau)}.$$

Note that the impulse response gives the system transfer function directly as output; for an impulse at the input the output is equal to the transfer function times 1. This is why it is a useful function. However, it is not always possible, or desirable, to stimulate a system with this kind of input.

Taking the inverse Laplace transform of the above, we get

$$y(t) = \frac{K}{\tau} \exp\left(\frac{-t}{\tau}\right).$$

The response of this system is shown in Figure 3.8. Notice that at $t = 0$ the response is K/τ. This can be plotted directly from Scilab using the following code:

```
s=%s; // Declare 's' as Laplace operator
K = 1; // Assume K=1
tau = 1; // Assume tau = 1
sys = syslin('c',K/(tau*s+1)); // Create system transfer function
t = 0:0.01:6; // Create time vector
y = csim('impulse',t,sys); // Create output data
plot(t,y);
```

For Matlab, use:

```
s=tf('s'); % Declare 's' as Laplace operator
K = 1; % Assume K=1
tau = 1; % Assume tau = 1
sys = K/(tau*s+1); % Create system transfer function
impulse(sys);
```

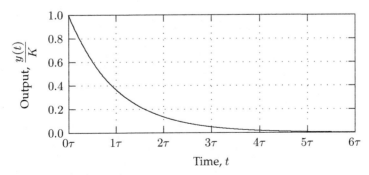

FIGURE 3.8 Impulse response of a first-order system.

3.4.2 Step response

Often it is not possible to excite a system with an impulse, since the impulse must have a large amplitude and short time duration. Exciting the system with a step signal may be more realistic and closer to what the real system would experience. For example, when switching on an electric motor, the (constant) voltage applied to the terminals will be applied almost instantaneously. Similarly, when opening a valve in a tank system for a constant inflow, the time taken to open the valve may be negligible compared with the dynamics of the tank.

The Laplace transform of the unit step is $\frac{1}{s}$, so the transfer function for the output becomes

$$\frac{Y(s)}{1/s} = \frac{K}{(\tau s + 1)} \Rightarrow Y(s) = K\left(\frac{1}{s(\tau s + 1)}\right).$$

Using partial fractions, we get:

$$Y(s) = K\left(\frac{1}{s} - \frac{\tau}{(\tau s + 1)}\right).$$

Taking the inverse Laplace transform:

$$y(t) = K\left[1 - \exp\left(\frac{-t}{\tau}\right)\right].$$

The output from a first-order system when excited with a step input is shown in Figure 3.9. This can be plotted similarly as above, by replacing the "impulse" argument by "step."

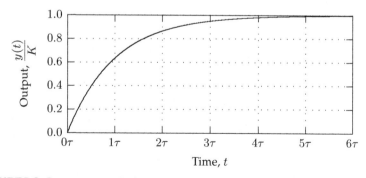

FIGURE 3.9 Step response of a first-order system.

3.4.3 Ramp response

The third type of input, the unit ramp, may in some cases better represent the actual input experienced by a system. The unit ramp is defined as

$$
\begin{cases}
u(t) = 0 \ \text{if } t < 0 \\
u(t) = t \ \text{if } t \geq 0
\end{cases}
$$

and its Laplace transform is $1/s^2$. For a first-order system excited by a ramp input, the output is

$$
Y(s) = \frac{K}{s^2(\tau s + 1)}.
$$

Using partial fractions, we have

$$
\frac{K}{s^2(\tau s + 1)} = \frac{A}{s^2} + \frac{B}{s} + \frac{C}{(\tau s + 1)},
$$

and we need to find the values for A, B, and C. Multiplying out and equating for orders of s, we get

$$
K = A(\tau s + 1) + Bs(\tau s + 1) + Cs^2,
$$

$$
\begin{aligned}
s^0 &: K = A &&\Rightarrow A = K, \\
s^1 &: 0 = A\tau + B &&\Rightarrow B = -K\tau, \\
s^2 &: 0 = B\tau + C = -K\tau^2 + C &&\Rightarrow C = K\tau^2.
\end{aligned}
$$

Hence, the output is

$$
Y(s) = K\left(\frac{1}{s^2} - \frac{\tau}{s} + \frac{\tau^2}{(\tau s + 1)}\right).
$$

Taking the inverse Laplace transform we get the time domain solution:

$$
y(t) = K\left[t - \tau + \tau \exp\left(-\frac{t}{\tau}\right)\right].
$$

Figure 3.10 shows the first-order system response to a unit ramp input. The response starts off with an exponential growth but after a time the response follows the input with an offset of $K\tau$. This offset will always be present; the output cannot track the input. (Compare this with the step response above; the input was one and the output eventually also settled at one.)

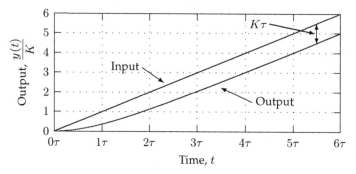

FIGURE 3.10 Ramp response for a first-order system.

3.4.4 Influence of pole placement on system response

A first-order system in the standard form has one pole and no zeros. The pole is always real (it has no imaginary part), because the characteristic equation $\tau s + 1 = 0$ always has a real solution.

Section 2.5.1 showed that the only thing influencing the response of a first-order system is the gain, K, and the time constant, τ. (This can also be seen from the general solution.) The gain, K is just a gain factor; it does not influence the *shape* of the reponse. The time constant τ will be associated with the system pole, since it's part of the characteristic equation.

Figure 3.11 shows the characteristics of a first-order system for three values of the time constant, with $K = 1$. Figure 3.11a shows the pole placement and Figure 3.11b shows the corresponding time response to a step change in the input.

From the time response plots, notice that the slope at the origin is always equal to K/τ. After a period of one time constant, $t = \tau$, the output is at 63.2% of its final value, after a period of 3τ the output is at 95% of its final value, and after 5τ the output is at 99% of its final value.

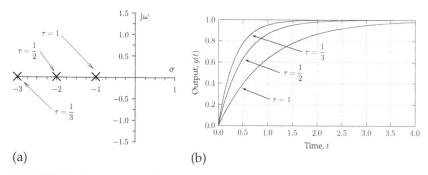

(a) (b)

FIGURE 3.11 Pole placement and corresponding step response for first-order systems. (a) Pole-zero map. (b) Time response.

3.5 TRANSIENT RESPONSE OF SECOND-ORDER SYSTEMS

The standard form for a second-order system differential equation is:

$$\frac{d^2y}{dt^2} + 2\zeta\omega_n\frac{dy}{dt} + \omega_n^2 y = K\omega_n^2 u(t),$$

and the transfer function is

$$\frac{Y(s)}{U(s)} = \frac{K\omega_n^2}{s^2 + 2\zeta\omega_n s + \omega_n^2}.$$

The value of ζ is called the damping ratio and there are three main system responses governed by the damping ratio:

i. Underdamped $0 \le \zeta \le 1$,
ii. Critically damped $\zeta = 1$,
iii. Overdamped $\zeta > 1$.

The characteristic equation is given by $s^2 + 2\zeta\omega_n s + \omega_n^2 = 0$ and the roots of the characteristic equation can be determined by the equation $r_{1,2} = \dfrac{-b \pm \sqrt{b^2 - 4ac}}{2a}$. Hence, we have

$$r_{1,2} = \frac{-2\zeta\omega_n \pm \sqrt{4\zeta^2\omega_n^2 - 4\omega_n^2}}{2},$$

$$r_{1,2} = \frac{-2\zeta\omega_n \pm 2\omega_n\sqrt{\zeta^2 - 1}}{2},$$

$$r_{1,2} = -\zeta\omega_n \pm \omega_n\sqrt{\zeta^2 - 1}.$$

The roots are different in nature depending on the value of ζ:

i. $0 \le \zeta \le 1$ (underdamped): roots are complex conjugate.
ii. $\zeta = 1$ (critically damped): roots are real (and the same).
iii. $\zeta > 1$ (overdamped): roots are real (and different).

The pole zero plots in Figure 3.12a-c show the location of the poles (i.e., the roots of the characteristic equation) for different values of damping ratio: overdamped, $\zeta = 1.5$; critically damped, $\zeta = 1$; and underdamped, $\zeta = 0.5$. Note that in the underdamped case, there is a component on the imaginary (vertical) axis meaning that the response has an oscillatory component.

3.5.1 Impulse response

The Laplace transform of an impulse input is 1, so the output $Y(s)$ is simply the system transfer function itself. Hence, the output for the second-order system is (assuming $K = 1$):

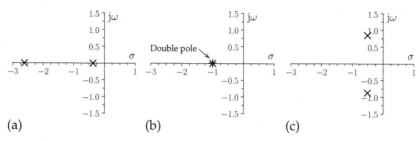

FIGURE 3.12 Influence of damping ratio on second order system poles. (a) Overdamped ($\zeta = 1.5$). (b) Critically damped ($\zeta = 1$). (c) Underdamped ($\zeta = 0.5$).

$$Y(s) = \frac{\omega_n^2}{s^2 + 2\zeta\omega_n s + \omega_n^2}$$

The inverse Laplace transform depends on the value of ζ. We will simply state the results without showing the working (which is somewhat laborious).

$$0 < \zeta < 1 \quad y(t) = \frac{\omega_n}{\sqrt{1 - \zeta^2}} e^{-\zeta\omega_n t} \sin\left(\omega_n\sqrt{1 - \zeta^2}t\right),$$

$$\zeta = 1 \quad y(t) = \omega_n^2 t e^{-\omega_n t},$$

$$\zeta > 1 \quad y(t) = \frac{\omega_n}{2\sqrt{\zeta^2 - 1}} \left\{ \exp\left[-\left(\zeta - \sqrt{\zeta^2 - 1}\right)\omega_n t\right] - \exp\left[-\left(\zeta + \sqrt{\zeta^2 - 1}\right)\omega_n t\right] \right\}$$

These responses are shown in Figure 3.13. For the over- and critically damped cases ($\zeta \geq 1$) the response is not oscillatory. For the underdamped case, the actual frequency of oscillation is given by the *damped natural frequency*, ω_d:

$$\omega_d = \omega_n\sqrt{1 - \zeta^2}, \quad \zeta < 1.$$

This can be plotted directly in Scilab using the following code:

```
s=%s; // Declare 's' as Laplace operator
omegaN = 1; // Assume undamped natural frequency is 1
for zeta = [0.2,0.5,1,2] // Loop over 4 zeta values
  sys = omegaN^2/(s^2+2*zeta*omegaN*s+omegaN^2); // Create TF
  t = 0:0.01:20; // Create time vector
  y = csim('impulse',t,sys); // Create output data
  plot(t,y);
end // End for loop
```

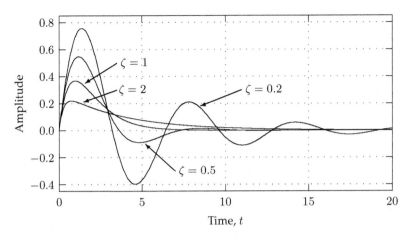

FIGURE 3.13 Impulse response of a second-order system.

The equivalent Matlab code is:

```
s=tf('s'); % Declare 's' as Laplace operator
omegaN = 1; % Assume undamped natural frequency is 1
for zeta = [0.2,0.5,1,2] % Loop over 4 zeta values
  sys = omegaN^2/(s^2+2*zeta*omegaN*s+omegaN^2); % Create TF
  impulse(sys);
  hold on;
end % End for loop
```

3.5.2 Step response

If we excite the second-order system with a step input $U(s) = \dfrac{1}{s}$, we get (again assuming $K = 1$):

$$Y(s) = \frac{\omega_n^2}{s(s^2 + 2\zeta\omega_n s + \omega_n^2)}.$$

The time-domain solution to this equation changes depending on the range of ζ. Again we will simply state the results for each case of damping ratio.

For the underdamped case:

$$0 < \zeta < 1 : Y(s) = \frac{1}{s} - \frac{s + \zeta\omega_n}{(s + \zeta\omega_n)^2 + (1 - \zeta^2)\omega_n^2} - \frac{\zeta\omega_n}{(s + \zeta\omega_n)^2 + (1 - \zeta^2)\omega_n^2},$$

$$y(t) = 1 - \frac{e^{-\zeta\omega_n t}}{\sqrt{1 - \zeta^2}} \sin\left(\omega_d t + \tan^{-1}\frac{\sqrt{1 - \zeta^2}}{\zeta}\right),$$

$$\text{where } \omega_d = \left(\sqrt{1 - \zeta^2}\right)\omega_n.$$

For the critically damped case:

$$\zeta = 1 : Y(s) = \frac{\omega_n^2}{s(s + \omega_n)^2}$$

$$y(t) = 1 - e^{-\omega_n t}(1 + \omega_n t).$$

For the overdamped case:

$$\zeta > 1 : Y(s) = \frac{\omega_n^2}{s\left(s + \zeta\omega_n + \omega_n\sqrt{\zeta^2 - 1}\right)\left(s + \zeta\omega_n + \omega_n\sqrt{\zeta^2 - 1}\right)}$$

$$y(t) = 1 - \frac{\omega_n}{2\sqrt{\zeta^2 - 1}}\left\{\frac{\exp\left[-\left(\zeta + \sqrt{\zeta^2 - 1}\right)\omega_n t\right]}{\zeta + \sqrt{\zeta^2 - 1}}\right.$$

$$\left. - \frac{\exp\left[-\left(\zeta - \sqrt{\zeta^2 - 1}\right)\omega_n t\right]}{\zeta - \sqrt{\zeta^2 - 1}}\right\}$$

Again, this can be plotted directly from the Laplace domain similarly as above, by replacing "impulse" with "step" in the code. The output is shown in Figure 3.14. The same characteristics as above can be seen: with critical damping the system settles in the quickest manner without oscillations, for lower damping ratios we have overshoot and oscillations, whereas for higher values of damping it takes longer for the response to settle.

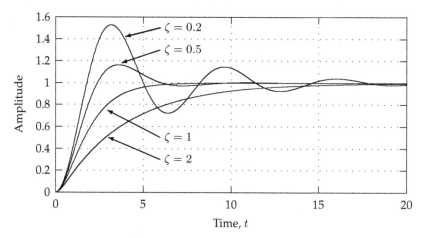

FIGURE 3.14 Step response of a second-order system.

3.5.3 Example: Pole placement in the mass-spring-damper system

We have seen that when the damping coefficient in a second-order system changes, the dynamics of the system are influenced. Obviously, the other system variables also influence the behavior of the system, but not necessarily in the same ways.

Let us go back to the spring-mass-damper system. Changing the damping, C, in that system will have a similar effect to that shown above for varying ζ, since C influences the first-order term in the characteristic equation. But what, for example, is the effect of changing the spring stiffness, k? We can try this using the simulation software with the following piece of code, in Scilab:

```
s=%s; // Declare 's' as Laplace operator
m = 1; // Mass (constant)
C = 2; // Damping (constant)
for k = [0.75,1,2] // Loop over some spring stiffness values
  sys = 1/(m*s^2+C*s+k); // Create TF
  t = 0:0.01:15; // Create time vector
  y = csim('step',t,sys); // Create output data
  scf(0); plot(t,y); // Plot time response
  scf(1); plzr(sys); // Plot pole-zero map
end
```

The same code for Matlab is:

```
s=tf('s'); % Declare 's' as Laplace operator
m = 1; % Mass (constant)
C = 2; % Damping (constant)
for k = [0.75,1,2] % Loop over some spring stiffness values
  sys = 1/(m*s^2+C*s+k); % Create TF
  figure(1); step(sys); hold on; % Plot time response
  figure(2); pzmap(sys); hold on; % Plot pole-zero map
end
```

The output of this should look something like that in Figure 3.15. These plots are for three values of k (0.75, 1, and 2) with the other variables held constant. In this case, we see that the dynamics change as k varies: from being underdamped for low k, the system goes to being critically damped for $k = 1$ and underdamped for higher k. For $k = 2$, the poles have imaginary parts and the system therefore has an oscillatory response; this can be seen as a small overshoot in the time response plot.

Moreover, the final value also depends on the value of k. We can double-check this using the final value theorem (see page 39). Remember that the system transfer function is $G(s) = \dfrac{1}{ms^2 + Cs + k}$, and the (step) input is $U(s) = \dfrac{1}{s}$. Putting this in to the final value theorem, we get:

$$\lim_{t \to \infty} f(t) = \lim_{s \to 0} sF(s) = \lim_{s \to 0} s \cdot \frac{1}{ms^2 + Cs + k} \cdot \frac{1}{s} = \frac{1}{k}.$$

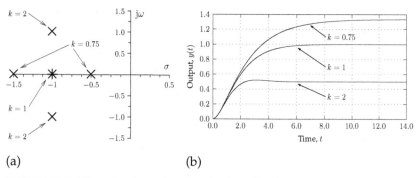

(a)   (b)

FIGURE 3.15 Effect of spring constant on the dynamics of the mass-spring-damper system. (a) Pole-zero map. (b) Time response.

Comparing with the final values in the time plot, this looks correct. Hence, we see that varying the spring constant influences both the type of response from the system and the amplitude of the output. You can do a similar analysis to study the effect of changing the mass.

3.6 HIGHER-ORDER SYSTEMS

As we have seen, physical systems are often of first or second order. However, we will frequently have systems of higher order, for example when several physical components—each having its own dynamics associated with it—are interconnected.

Consider, for example, a ship rudder system consisting of the rudder itself and the steering gear. The rudder may be modeled as an overdamped second-order system, with the input being the torque applied by the steering gear and the output being rudder position. The steering gear will have its own dynamics, possibly of first order, translating an input demand signal into torque applied on the rudder. In block diagram form, this can be represented as shown in Figure 3.16.

The total system, i.e., from demand signal to rudder position, is now of third order, and has the transfer function

$$\frac{Y(s)}{U(s)} = \frac{1}{(\tau s + 1)(s + a)(s + b)}.$$

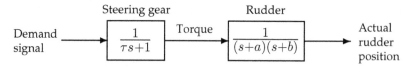

FIGURE 3.16 Third-order dynamic system.

Similarly, if there are other components in the system, their dynamic properties must be taken into account when modeling the system. For example, if it is the ship heading we are interested in, we would need a further block to represent the dynamics associated with the response of the ship to a change in the rudder position. Another example which often comes up in practice is the dynamics associated with sensors.

3.6.1 Dominant poles

Nevertheless, we don't always have to consider every single component of a system which could influence the system dynamics. Consider, for example, the system model:

$$G = \frac{1}{(s + 10)} \frac{1}{(s^2 + 2s + 17)}.$$

This system has three poles, one at -10 and two complex conjugate poles at -1. These poles determine the time response of this system: the speed of response is determined by the real value and the transient response characteristics depend on the complex values. The dynamics associated with the pole at -10 will be much faster than those resulting from the other two poles, and we can therefore expect the two second-order poles to dominate the reponse.

The first-order term has a gain of 0.1; the dynamics are determined by the time constant $\tau = 0.2$. (We find these by putting the expression in standard form.) We can try and remove the dynamics of the first-order element (the s terms) and simply approximate it as a simple gain. The overall system transfer function becomes:

$$G \simeq \frac{0.1}{(s^2 + 2s + 17)}.$$

Figure 3.17 shows a plot the original and simplified systems response to a step input, we see that the simplified, second-order model provides a reasonable approximation of the third-order system in this case. We can conclude that it is the poles closest to the imaginary axis in the Argand diagram that have the highest influence on the system response.

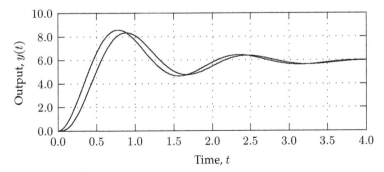

FIGURE 3.17 Dominant poles and effect of simplification.

3.7 ELECTRIC PUMP DRIVE IN THE s-DOMAIN

In Chapter 2 we modeled an electric pump drive by the following equations:

$$iR + L\frac{di}{dt} = V - k_e\frac{d\theta}{dt},$$

$$J\frac{d^2\theta}{dt^2} + k_f\frac{d\theta}{dt} = T_e - T_L, \quad \text{where}$$

$$T_e = k_a \cdot i.$$

and the inputs to the system are the voltage applied to the motor terminals, V, and the load torque T_L. We can solve these equations for the motor output as a function of the inputs. Rotational speed, $\frac{d\theta}{dt}$, will be used as the output, as we will later develop controllers for motor speed.

ELECTRICAL MODEL

Based on the voltage drops through the electric circuit, we developed the following equation:

$$iR + L\frac{di}{dt} = V - k_e\frac{d\theta}{dt},$$

where $k_e\frac{d\theta}{dt}$ is the back emf generated in the motor windings. Taking the Laplace transform of this equation with zero initial conditions we get

$$RI(s) + LsI(s) = V - k_e s\theta(s).$$

Rearranging this:

$$I(s) = \frac{1}{(Ls + R)}(V - k_e s\theta(s)).$$

This is a transfer function with current, I, as the output and voltage input minus back emf as the input. Taking the Laplace transform of the third equation from above, the electromagnetic torque can be found as:

$$T_e = k_a I(s).$$

These relationships can be illustrated in block diagram form as shown in Figure 3.18.

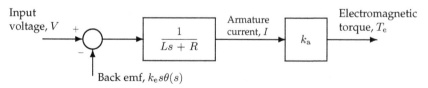

Back emf, $k_e s\theta(s)$

FIGURE 3.18 Electrical submodel in block diagram form.

MECHANICAL MODEL

The equation for the mechanics of the motor and impeller assembly was

$$J\frac{d^2\theta}{dt^2} + k_f\frac{d\theta}{dt} = T_e - T_L.$$

Taking Laplace transforms with zero initial conditions we get

$$Js^2\theta(s) + k_f s\theta(s) = T_e - T_L.$$

Rearranging for rotational speed:

$$s\theta = \frac{1}{(Js + k_f)}(T_e - T_L)$$

This is a transfer function with output $s\theta$ and input $T_e - T_L$. Note that the output, $s\theta$, is equivalent to rotational velocity $\frac{d\theta}{dt}$. In block diagram form, we have a model as shown in Figure 3.19

The load torque T_L will depend on external conditions (e.g., fluid properties, pressures, etc.) which could vary, however for this example let us assume that the pump works under stable conditions and that the load torque is approximately proportional to the rotating speed of the pump so that:

$$T_L = k_L\frac{d\theta}{dt},$$

where k_L is a load factor. In the Laplace domain, this is:

$$T_L = k_L s\theta(s).$$

The block diagram model can be adapted to include this by simply feeding back the rotational speed and multiplying it with the load factor, as shown in Figure 3.20.

TOTAL MODEL

Combining the equations and rearranging, the total transfer function for the output (rotational speed) as a function of the inputs (voltage and load torque) can be expressed as:

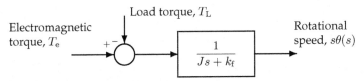

FIGURE 3.19 Mechanical submodel in block diagram form.

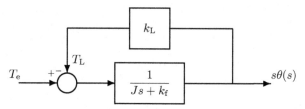

FIGURE 3.20 Modified mechanical submodel.

$$\frac{s\theta(s)}{V(s)} = \frac{k_a}{JLs^2 + (JR + k_fL + k_LL)s + (k_fR + k_LR + k_ak_e)}.$$

The two block diagram models can also be joined. Conveniently, the output from the "electric" block diagram is the input to the "mechanical," so combining them is straightforward. The back emf which is required is simply the output speed multiplied with the motor constant, so the $s\theta$ signal can be sourced at the output, multiplied with the constant, and fed it back to the input on the left-hand side. Figure 3.21 shows the total block diagram model.

SIMULATING PUMP DRIVE DYNAMIC RESPONSE
Suppose we have a system with coefficients:
- $L = 0.4\,\text{H}$
- $R = 1\,\Omega$
- $k_a = 0.5\,\text{N}\,\text{m}/\text{A}$
- $k_L = 1\,\text{N}\,\text{m}\,\text{s}$
- $J = 0.2\,\text{kg}\,\text{m}^2/\text{s}^2$
- $k_f = 0.2\,\text{N}\,\text{m}\,\text{s}$
- $k_e = 0.5\,\text{V}\,\text{s}$

We can now simulate the response of the pump drive to an excitation with a value of, e.g., 10 V in the input. In Scilab, the following code will do this:

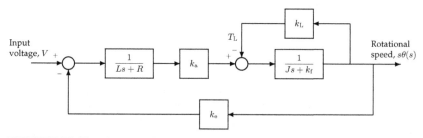

FIGURE 3.21 Electric pump drive total model block diagram.

```
s=%s;
// Motor coefficients:
  L = 0.4;
  R = 1;
  ka = 0.5;
  kL = 1;
  J = 0.2;
  kf = 0.2;
  ke = 0.5;
// Motor transfer function:
  Gmotor = syslin('c',ka/(J*L*s^2 + (J*R+kf*L+kL*L)*s +
    (kf*R+kL*R+ka*ke)));
// Simulate response for a step input of 10 and plot
    output:
  t = 0:0.01:3;
  y = 10*csim('step',t,Gmotor);
  plot(t,y);
```

The equivalent code for Matlab is:

```
s=tf('s');
% Motor coefficients:
  L = 0.4;
  R = 1;
  ka = 0.5;
  kL = 1;
  J = 0.2;
  kf = 0.2;
  ke = 0.5;
% Motor transfer function:
  Gmotor = ka/(J*L*s^2 + (J*R+kf*L+kL*L)*s + (kf*R+kL*R+
    ka*ke));
% Simulate response for a step input of 10 and plot output:
  step(10*Gmotor);
```

Figure 3.22 shows the response to the step input. We can see that for a demand input of 10, the output is between 3 and 4.

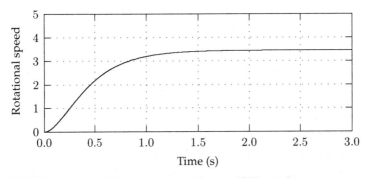

FIGURE 3.22 Response of the electric pump drive to a 10 V input change.

QUESTIONS

3.1 For the mass-spring system from Chapter 2, we have $\dfrac{d^2y}{dt^2} + \dfrac{k}{m}y = u(t)$, where $u(t)$ is the input force. Find the transfer function in terms of s and show how one can get the natural frequency of the system. (Assume that the initial conditions are zero).

3.2 A second-order system has a natural frequency of $4\,\text{rad/s}$ and is over-damped with a damping ratio of 1.25. Find the transfer function and put it into partial fraction form.

3.3 For the tank system shown, the differential equation is $\dfrac{dh}{dt} + \dfrac{kh}{A} = \dfrac{Q_{in}}{A}$ (see page 19), where Q_{in} is inflow rate, A is the surface area of the fluid, and k is the proportional constant relating fluid level, h, to outflow rate Q_{out}.

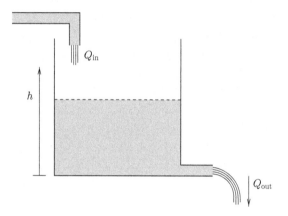

(a) Find the transfer function, relating Q_{in} to h, assuming that the tank starts off empty.

(b) Using the Laplace transforms, determine the time domain response to a unit step and a unit ramp input.

3.4 A system is characterized by the differential equation

$$\frac{1}{B}\frac{d^2y}{dt^2} + A\frac{dy}{dt} + By(t) = 7u(t),$$

where u is the input, y is the output, and A and B are constants. Find the transfer function, assuming all initial conditions are zero, and describe the characteristics of the system response if $A = 1$ and $B = 7$. Find a value of A to ensure that the response is not oscillatory.

3.5 For the mass-spring-damper system, determine values for the spring constant and damping coefficient which will cause a mass of $2\,\text{kg}$ to have an overdamped response ($\zeta = 2$) and a undamped natural frequency of $3\,\text{rad/s}$. Will the mass oscillate at $3\,\text{rad/s}$?

3.6 Match the transfer function to the unit step responses:

TF1: $\dfrac{1}{0.2s + 1}$ TF4: $\dfrac{10}{s^2 + s + 10}$

TF2: $\dfrac{10}{s^2 + s + 30}$ TF5: $\dfrac{10}{s^2 + 10s + 30}$

TF3: $\dfrac{1}{s + 1}$ TF6: $\dfrac{1}{s + 2}$

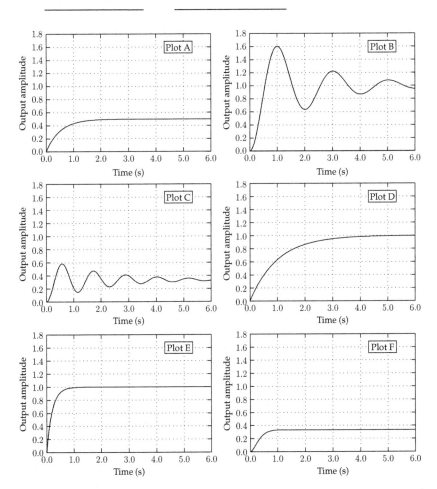

Chapter | Four

Feedback Control

CHAPTER POINTS

- Block diagram reduction
- Error actuated feedback control
- Closed-loop systems dynamics
- Controller actions: proportional, integral, derivative, velocity feedback
- Disturbance rejection
- Examples: electric pump drive control and ship autopilot system

4.1 BLOCK DIAGRAM REDUCTION

The last chapter showed how systems can be represented in block diagrams. Such block diagrams can become quite complex, for example, if the system has internal feedback loops like we saw for the electric motor in the pump drive example. Sometimes we therefore need to reduce a complex system diagram into a more manageable form. We can do this mathematically by looking at the algebra of the transfer function equations of each individual block, but often it is easier just to manipulate the blocks directly. Table 4.1 shows some useful block diagram reduction rules.

4.1.1 Block diagram reduction example

Consider the system block diagram shown below. Let us try to simplify this using the rules described above, in order to get it into a more manageable form.

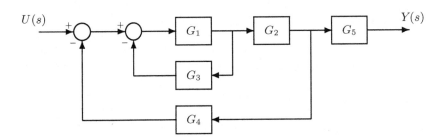

Marine Systems Identification, Modeling, and Control. http://dx.doi.org/10.1016/B978-0-08-099996-8.00004-5

TABLE 4.1 Block Diagram Reduction Rules

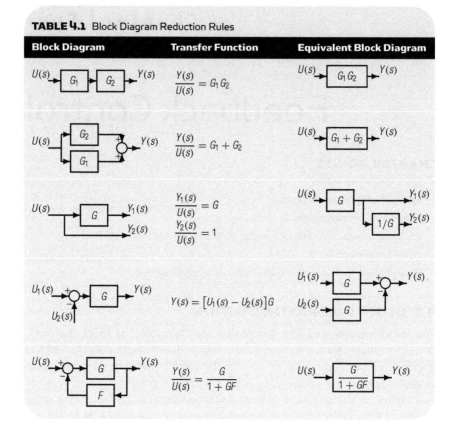

Block Diagram	Transfer Function	Equivalent Block Diagram
$U(s) \to \boxed{G_1} \to \boxed{G_2} \to Y(s)$	$\dfrac{Y(s)}{U(s)} = G_1 G_2$	$U(s) \to \boxed{G_1 G_2} \to Y(s)$
$U(s) \to \boxed{G_2},\ \boxed{G_1} \to Y(s)$	$\dfrac{Y(s)}{U(s)} = G_1 + G_2$	$U(s) \to \boxed{G_1 + G_2} \to Y(s)$
$U(s) \to \boxed{G} \to Y_1(s),\ Y_2(s)$	$\dfrac{Y_1(s)}{U(s)} = G$ \quad $\dfrac{Y_2(s)}{U(s)} = 1$	$U(s) \to \boxed{G} \to Y_1(s);\ \boxed{1/G} \to Y_2(s)$
$U_1(s) \to \oplus \to \boxed{G} \to Y(s),\ U_2(s)$	$Y(s) = [U_1(s) - U_2(s)]G$	$U_1(s) \to \boxed{G} \to \oplus \to Y(s);\ U_2(s) \to \boxed{G}$
$U(s) \to \oplus \to \boxed{G} \to Y(s),\ \boxed{F}$	$\dfrac{Y(s)}{U(s)} = \dfrac{G}{1 + GF}$	$U(s) \to \boxed{\dfrac{G}{1 + GF}} \to Y(s)$

We use the last rule to simplify the inner feedback loop:

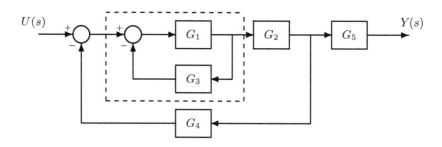

Having replaced this with the equivalent, single block, transfer function (see below), we can use the third rule from above to move the take-off point over the last block, G_5:

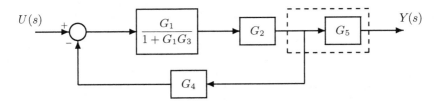

The take-off point is moved across the block by dividing by G_4 in the feedback loop. Now the three blocks in the forward loop can be combined together (using the first rule):

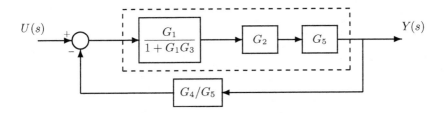

With this we get a standard feedback block diagram:

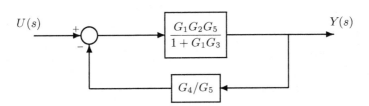

We can now use the rules to derive a single transfer function describing the complete system. Using the last rule from Table 4.1, we get:

$$\frac{Y(s)}{U(s)} = \frac{\dfrac{G_1 G_2 G_5}{1 + G_1 G_3}}{1 + \dfrac{G_4}{G_5} \cdot \dfrac{G_1 G_2 G_5}{1 + G_1 G_3}}$$

$$= \frac{G_1 G_2 G_5}{1 + G_1 G_3 + \dfrac{G_4}{G_5} G_1 G_2 G_5}$$

$$= \frac{G_1 G_2 G_5}{1 + G_1 G_3 + 1 + G_1 G_2 G_4}.$$

4.2 ERROR-ACTUATED FEEDBACK CONTROL

Figure 4.1 shows a model of a system or plant, $G(s)$, with input $U(s)$ and output $Y(s)$. The plant may not perform as we would like and may be subject to disturbances (e.g., the speed of a motor will be sensitive to load changes). These disturbances may be difficult or impossible to control and in fact we may not even know what they are. We would therefore like to be able to modify the plant input (e.g., the fueling rate in a diesel genset) in order to keep the output (e.g., engine speed) at the desired value.

As discussed in Chapter 1, this is achieved by using a sensor to monitor the output, and a controller, $C(s)$, to adjust the plant input according to the *error signal*, e, i.e., the difference between the actual output and the desired value. The sensor, or feedback loop, may also have dynamics associated with it, and we can represent these with a transfer function $H(s)$.

The use of such error-actuated feedback control loops may influence the system in a number of ways, including the steady-state response, its dynamics and speed of response, and even the stability of the system. The rest of this chapter and the next chapter will discuss these aspects in detail.

4.2.1 Steady-state tracking and disturbance rejection

In terms of the steady-state response of a system, there are two objectives for using feedback control. We want to be able to *track* any changes in the reference input $R(s)$, i.e., that the output follows the demand, and also *reject* any disturbances $D(s)$ to the plant, over which we usually have no control. The relative importance of these will vary according to the plant; for example, a diesel engine generator set may have a fixed setpoint which does not change (e.g., 1500 rpm) and the control system maintains that speed despite any variations in the disturbance (the electric load). In other plants, the tracking may be the key aspect, with disturbances being less influential, for example, controlling a pod drive to provide the propulsive thrust demanded at any time. There may also be a combination, such as a ship autopilot system, in which the setpoint frequently changes and there are also significant disturbances.

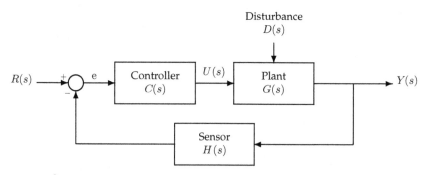

FIGURE 4.1 Error-actuated feedback control system.

4.2.2 Effect on dynamics of closing the loop

Consider an open-loop tank-heating system as shown in Figure 4.2, in which the temperature of a tank is regulated by flowing steam through a heat exchanger in the tank. A steam valve controls the flow based on an input signal, R, which is increased by a gain K to set the valve position. The output is the temperature in the tank, Y.

We can assume that the tank temperature response to a change in steam flow rate can be expressed as a first-order transfer function. This situation can then be illustrated in block diagram form as shown in Figure 4.3.

The transfer function for this system is $\dfrac{Y(s)}{R(s)} = \dfrac{K}{1 + \tau s}$. The effect of varying the gain K in this system is simply a variation in the *magnitude* of the output in response to some change in the demand signal $R(s)$. (See Section 2.5.1 for the effect of varying gain in first-order systems.) The *shape* of the response curve and the speed of response are not influenced. Figure 4.4 shows the time reponse to a step input for this system with $K = 1$ and $\tau = 1$.

What will happen if we "close the loop" by making it a feedback system? This situation is shown in Figure 4.5. In this case, the demand signal to the valve is not the input directly, but instead the difference between the input R and the actual temperature Y, i.e., the error signal e. The closed-loop system in block diagram form is shown in Figure 4.6.

From the block diagram reduction rules, we known that the transfer function for a closed-loop system is

$$\frac{Y(s)}{R(s)} = \frac{G(s)}{1 + G(s)H(s)} = \frac{\frac{K}{1+s\tau}}{1 + \frac{K}{1+s\tau}} = \frac{K}{\tau s + 1 + K}.$$

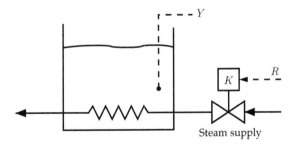

FIGURE 4.2 Tank-heating system.

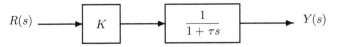

FIGURE 4.3 Open-loop tank temperature control.

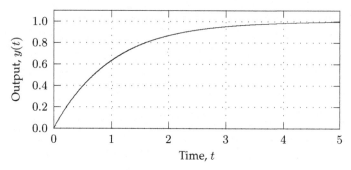

FIGURE 4.4 Open-loop time response for tank-heating system.

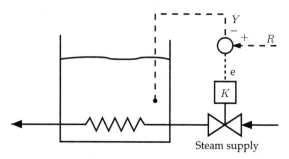

FIGURE 4.5 Closed-loop tank-heating system.

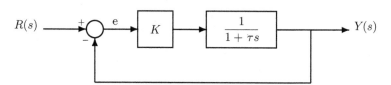

FIGURE 4.6 Closed-loop system transfer function.

This system has a pole at $s = \dfrac{-1}{\tau} - \dfrac{K}{\tau}$. The open-loop system pole was $s = \dfrac{-1}{\tau}$; hence, the pole has been shifted by $-\dfrac{K}{\tau}$, making the system more dynamically stable (since the pole is moving toward the left in the Argand diagram). We also know from the previous chapter that poles to the left have faster dynamics, so the feedback loop has influenced the responsiveness of the system as well.

Let us compare the open- and closed-loop response to a step input ($R(s) = 1/s$). The open-loop response was shown in the plot above. For the closed-loop case, we have

$$Y(s) = G(s) \cdot R(s) = \frac{K}{\tau s + 1 + K} \cdot \frac{1}{s}.$$

The time response can be plotted for some values of K with the following Scilab code:

```
s=%s;
tau = 1;
for K = [1,2,5,50];
  sys = syslin('c',K/(1+tau*s+K));
  t = 0:0.01:2;
  y = csim('step',t,sys);
  plot(t,y);
end
```

The equivalent code for Matlab is:

```
s=tf('s');
tau = 1;
hold on;
for K = [1,2,5,50];
  sys = K/(1+tau*s+K);
  t = 0:0.01:2;
  step(sys,t);
end
```

The result is shown in Figure 4.7; it can be seen that closing the loop significantly increases the responsiveness of the system. The closed-loop time constant is $\dfrac{\tau}{1+K}$, which is shorter compared to the open-loop time constant, τ. Closing the loop therefore makes the system respond faster, and the speed of response increases with increasing K.

It can be seen that the system gain is decreased compared to the open loop, but the larger the value of K, the closer the gain approaches unity (which was the open loop gain). We can use the final value theorem to find the final value for any value of K:

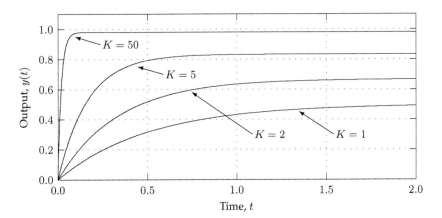

FIGURE 4.7 Closed-loop time response for varying K.

$$\lim_{t\to\infty} y(t) = \lim_{s\to0} s \cdot \frac{K}{\tau s + 1 + K} \cdot \frac{1}{s} = \frac{K}{1 + K}.$$

We see that introducing a feedback loop can improve the responsiveness of the system significantly, but with an influence on the system gain. By carefully designing the components we add to the feedback loop, we can obtain a better performance than with a simple open-loop system. It is, however, important to note that there will always be practical limits to the performance of a system; we cannot increase the gain indefinitely, since the physical components in the system will have limitations. (In this case, the maximum steam flow rate or valve opening.)

4.3 CONTROLLER ACTION: PROPORTIONAL, DERIVATIVE, INTEGRAL, VELOCITY FEEDBACK

We will now look at different types of "controller action" and their advantages and disadvantages. There are four types of controller action that we will analyze in terms of their effects on a second-order system (plant). These are
- proportional,
- integral,
- derivative, and
- velocity feedback.

We will look at their influence on a second-order plant, however, in general, the characteristics of using a certain type of control will be similar regardless of the type of plant.

4.3.1 Proportional control action

With this kind of control the output from the controller is equal to the input to the controller (i.e., the error signal) multiplied by a constant gain term, which we can adjust as part of tuning the control system. (As in the tank heater example above.) Figure 4.8 shows a closed-loop control system with a proportional controller for a second-order plant.

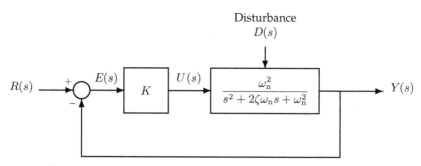

FIGURE 4.8 Second-order plant with proportional control.

Using the same method as above, and ignoring the disturbance $D(s)$ for now, we can derive the closed-loop transfer function for the system:

$$\frac{Y(s)}{R(s)} = \frac{C(s)G(s)}{1 + C(s)G(s)} = \frac{K\omega_n^2}{s^2 + 2\zeta\omega_n s + \omega_n^2(1 + K)} \ .$$

This is the transfer function for the whole system (plant and controller). Note that the characteristic equation of the closed-loop transfer function is second order, $s^2 + 2\zeta\omega_n s + \omega_n^2(1 + K) = 0$. The original plant was also second order; hence, the proportional control element has not changed the order of the system. However it has had the effect of "stiffening" the response because the constant term of the characteristic equation is changed. This is the same effect as changing the spring constant in the mass-spring-damper system.

Let us look at the steady-state response to a unit step in the reference input. We have

$$Y(s) = \frac{K\omega_n^2}{s^2 + 2\zeta\omega_n s + \omega_n^2(1 + K)} \cdot R(s),$$

where $R(s)$ is a unit step, $\frac{1}{s}$. Applying the final value theorem, we get

$$\lim_{t\to\infty} y(t) = \lim_{s\to 0} sY(s) = s\frac{K\omega_n^2}{s^2 + 2\zeta\omega_n s + \omega_n^2(1 + K)} \frac{1}{s} = \frac{K}{1 + K}.$$

The input was 1 (remember that this is the demanded value) and the steady-state output is $\frac{K}{1 + K}$, so the steady-state error is $r(\infty) - y(\infty) = 1 - \frac{K}{1 + K} = \frac{1}{1 + K}$. Hence, the proportional controller is unable to track the demand; there will always be a steady-state error.

We see that the larger we make K, the smaller the steady-state error becomes. However, we cannot make K too large since control systems rely on actuators to transfer energy from the controller to the plant; these actuators always have output limits (they saturate). In addition, an actuator with a greater range or faster response may be more expensive.

The performance of a proportional controller can also be studied using the simulation software. The following script defines the second-order system, the controller transfer function, and the total system, and plots the response to a step input for some values of K. The code for Scilab is:

```
s=%s;
zeta = 1; // Damping ratio
omegaN = 1; // Natural frequency
G = syslin('c',omegaN/(s^2+2*zeta*omegaN*s+omegaN^2));
    // System TF
t = 0:0.01:7;
```

```
Gunity=syslin('c',1,1);

for K = [1,2,5,10];
  C = K;
  Gclosed = (C*G)/.Gunity;
  y = csim('step',t,Gclosed);
  plot(t,y);
end
```

And the equivalent code for Matlab is:

```
s=tf('s');
zeta = 1; % Damping ratio
omegaN = 1; % Natural frequency
G = omegaN/(s^2+2*zeta*omegaN*s+omegaN^2); % System TF
t = 0:0.01:7;

hold on;
for K = [1,2,5,10];
  C = K;
  Gclosed = feedback(C*G,1);
  step(Gclosed,t);
end
```

Figure 4.9 shows the simulation output. From the plot we can see the steady-state error and how it changes depending on K. One further noticeable aspect is that the use of a proportional controller introduces oscillations in the response for higher values of K. This may be counter-intuitive at first sight, considering that we used a damping ratio (ζ) of 1, however it is important to be aware that the closed-loop system will have its own natural frequency and damping ratio. It can be shown that the closed-loop damping ratio is $\zeta_{cl} = \dfrac{\zeta}{\sqrt{1+K}}$. Using a proportional controller will therefore reduce the damping compared with the open-loop system, and the reduction depends on the value of K.

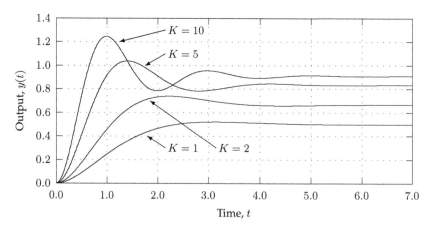

FIGURE 4.9 Time response for a second-order system with proportional control.

4.3.2 Integral control action

Recall from the Laplace transform table that the integration function in the s-domain is $\frac{1}{s}$. A pure integral controller consists of only the integrator term, with an adjustable controller gain, K. Figure 4.10 shows an integral controller acting on a second-order plant in block diagram form.

Ignoring the disturbance, the closed-loop transfer function, from reference input $R(s)$ to output $Y(s)$, is

$$\frac{Y(s)}{R(s)} = \frac{C(s)G(s)}{1 + C(s)G(s)} = \frac{K\omega_n^2}{s^3 + 2\zeta\omega_n^2 s^2 + \omega_n^2 s + K\omega_n^2}.$$

We see that the characteristic equation of the closed-loop transfer function is third order: $s^3 + 2\zeta\omega_n s^2 + \omega_n^2 s + K\omega_n^2 = 0$. The original plant was second order, but the integral control element has increased the order of the total system by 1. This means that the system has become more complex and may be more prone to instability.

Let's look at the steady-state response to a unit step. We have

$$Y(s) = \frac{K\omega_n^2}{s^3 + 2\zeta\omega_n s^2 + \omega_n^2 s + K\omega_n^2} R(s)$$

and $R(s) = \frac{1}{s}$. Applying the final value theorem:

$$\lim_{t\to\infty} y(t) = \lim_{s\to 0} sY(s) = s \cdot \frac{K\omega_n^2}{s^3 + 2\zeta\omega_n s^2 + \omega_n^2 s + K\omega_n^2} \cdot \frac{1}{s} = 1.$$

This means that for a demand in the reference input of 1, the output is 1. That is, there is no steady-state error, and the control system tracks the input (provided that the system is stable). The reason for this is that the integral controller acts not on the error itself like the proportional controller, but on the *aggregated* error. Therefore, as long as there is a non-zero error signal, the output from the controller will grow or decline until the error is removed.

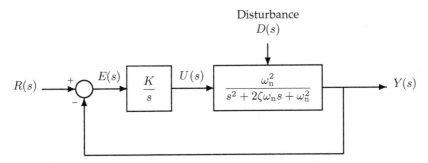

FIGURE 4.10 Second-order plant with integral control.

Let us simulate this system using the software and look at its dynamic properties. We use the same code as above, only changing the time range and the transfer function for the controller *C*.

For Scilab, we can use:

```
s=%s;
zeta = 1; // Damping ratio
omegaN = 1; // Natural frequency
G = syslin('c',omegaN/(s^2+2*zeta*omegaN*s+omegaN^2));
    // System TF
t = 0:0.01:30;
Gunity=syslin('c',1,1);

for K = [0.2,0.5,1];
  C = syslin('c',K/s);
  Gclosed = (C*G)/.Gunity;
  y = csim('step',t,Gclosed);
  plot(t,y);
end
```

The equivalent code for Matlab is:

```
s=tf('s');
zeta = 1; % Damping ratio
omegaN = 1; % Natural frequency
G = omegaN/(s^2+2*zeta*omegaN*s+omegaN^2); % System TF
t = 0:0.01:30;

hold on;
for K = [0.2,0.5,1];
  C = K/s;
  Gclosed = feedback(C*G,1);
  step(Gclosed,t);
end
```

Figure 4.11 shows the time response for the chosen values of controller gain. We can see that the output does go to 1 as time increases for all values of *K*, as predicted above. Further, increasing *K* gives a response which is faster initially but more oscillatory. Note, also, the significantly slower response compared with the proportional controller; this is a characteristic of the integral controller.

Another issue with this controller is that it will actually bring the system to instability if we increase the gain too much. (Try, e.g., with $K = 3$ in the simulation code.) So, in fact, by using integral control and thereby increasing the order of the plant, we have the possibility of the system going unstable if we increase *K* too much. We will discuss stability in more detail later and see why this happens.

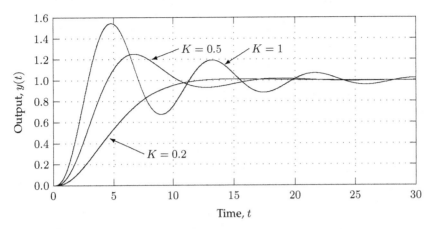

FIGURE 4.11 Time response for a second-order system with integral control.

4.3.3 Derivative control action

The derivative controller acts on the derivative, i.e., the rate of change, of the error signal. Recall from the Laplace transform table that the derivative function in the s-domain is s, and the controller gain is represented, as above, by K.

The control loop with a derivative controller is shown in Figure 4.12. The closed-loop transfer function is

$$\frac{Y(s)}{R(s)} = \frac{C(s)G(s)}{1 + C(s)G(s)} = \frac{K\omega_n^2 s}{s^2 + (2\zeta\omega_n + K\omega_n^2)s + \omega_n^2}.$$

We see that the characteristic equation of the closed-loop transfer function is second order: $s^2 + (2\zeta\omega_n + K\omega_n^2)s + \omega_n^2 = 0$. The original plant was also second order, but we see that a derivative control element has changed the first-order term. Hence, the closed-loop system has a damping ratio different to that of the open-loop system. (Note that in this case, the undamped natural frequency, ω_n is the same for the open- and closed-loop systems.)

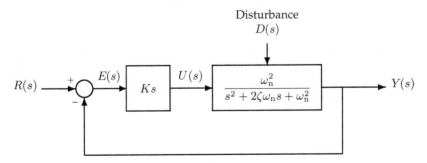

FIGURE 4.12 Second-order plant with derivative control.

If the closed-loop system has a characteristic equation $s^2 + 2\zeta_{cl}\omega_n s + \omega_n^2$, we have that

$$2\zeta_{cl}\omega_n = 2\zeta\omega_n + K\omega_n^2,$$

$$\zeta_{cl} = \zeta + \frac{K\omega_n}{2},$$

where ζ is the open-loop damping ratio and ζ_{cl} is that of the closed loop. Hence, the controller increases the damping as a function of the derivative gain value.

Let's look at the steady-state response to a unit step. As in the previous cases, we have

$$Y(s) = \frac{K\omega_n^2 s}{s^2 + (2\zeta\omega_n + K\omega_n^2)s + \omega_n^2} R(s)$$

and $R(s)$ is a unit step $\frac{1}{s}$. Applying the final value theorem:

$$\lim_{t\to\infty} y(t) = \lim_{s\to 0} sY(s) = s \cdot \frac{K\omega_n^2 s}{s^2 + (2\zeta\omega_n + K\omega_n^2)s + \omega_n^2} \cdot \frac{1}{s} = 0.$$

Since the input was a unit step (i.e., 1 in the steady state), the steady-state error is -1. Hence, derivative action on its own will not track the input, in fact it will not influence the steady-state response at all (only the dynamics of the system).

Also, if the error signal changes rapidly (perhaps due to signal noise or a step change in the input), the derivative of that signal will be large. (In fact, the derivative of a step change is infinite, but of course in real systems all signals will have some rise time, however small.) For this reason, derivative gain values in feedback control systems should be kept very small and we should never use derivative action alone.

4.3.4 Velocity feedback control

As an alternative method to modify the damping of a system we can use velocity feedback control. The controller structure in block diagram form is shown in Figure 4.13. With velocity feedback control, instead of modifying the error signal, we modify the feedback signal that generates the error signal, as shown in the figure.

The closed-loop transfer function is, as above, found using the block diagram reduction rules from the beginning of the chapter (note that we now have a $H(s)$ term).

$$\frac{Y(s)}{R(s)} = \frac{C(s)G(s)}{1 + C(s)G(s)H(s)} = \frac{\omega_n^2}{s^2 + (2\zeta\omega_n + K\omega_n^2)s + \omega_n^2}$$

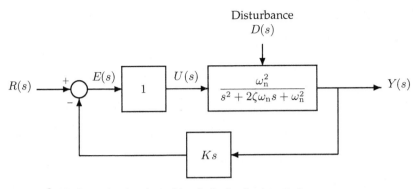

FIGURE 4.13 Second-order plant with velocity feedback control.

We see that the characteristic equation of the closed-loop transfer function is still second order: $s^2 + (2\zeta\omega_n + K\omega_n^2)s + \omega_n^2 = 0$. The velocity feedback control has changed the first-order term from $2\zeta\omega_n s$ to $2\zeta\omega_n s + K\omega_n^2 s$ in exactly the same way as the derivative control did. Again, the closed-loop system damping ratio, ζ_{cl}, is given by

$$\zeta_{cl} = \zeta + \frac{K\omega_n}{2}.$$

Hence, the damping ratio for the closed-loop controlled system is increased and is a function of the velocity feedback gain value.

Let us look at the steady-state response to a unit step. As before, we have

$$Y(s) = \frac{\omega_n^2}{s^2 + (2\zeta\omega_n + K\omega_n^2)s + \omega_n^2}R(s).$$

$R(s)$ is a unit step $\dfrac{1}{s}$. Applying the final value theorem:

$$\lim_{t\to\infty} y(t) = \lim_{s\to 0} sY(s) = s \cdot \frac{\omega_n^2}{s^2 + (2\zeta\omega_n + K\omega_n^2)s + \omega_n^2} \cdot \frac{1}{s} = 1.$$

That is, there is no steady-state error.

We see that the velocity feedback controller has a similar effect as the derivative controller, but with a couple of advantages. Firstly, the controller will track a change in the reference input as we have seen. Secondly, unlike in the derivative controller, rapid changes in the reference input (such as a step change) will not be amplified by the derivative term to produce very large controller outputs. Instead, the derivative gain acts on the plant output $Y(s)$.

4.3.5 Multiterm control

In practice, we often use more than one control term in a control system. We might use proportional and integral control together (PI), proportional and derivative (PD), or proportional, integral, and derivative (PID).

Considering the results above, we can see some immediate potential advantages of combining different controller terms. For example, the tendency of a proportional controller to give an oscillatory response at high gain values may be mitigated by adding derivative action, which, as we saw above, increases the damping. Similarly, adding integral action may get rid of the steady-state error in the proportional or derivative controllers.

Figure 4.14 shows the block diagram of a plant controlled by a PID (proportional-integral-derivative) controller. We are able to adjust the gains for each of the three controller terms individually, and this constitutes the (often challenging) task of tuning the controller. Using the block diagram reduction rules, we find that the controller transfer function can be written

$$C(s) = k_{\mathrm{p}} + \frac{k_{\mathrm{i}}}{s} + k_{\mathrm{d}}s.$$

Using the simulation software, we can study the effects of varying the controller coefficients. In Scilab, the following code will define this system:

```
s=%s;
zeta = 1; // Damping ratio
omegaN = 1; // Natural frequency
G = syslin('c',omegaN/(s^2+2*zeta*omegaN*s+omegaN^2));
    // System TF
t = 0:0.01:10;
Gunity=syslin('c',1,1);
Ki=1; // Integral gain
Kd=0; // Derivative gain
for Kp = [2,5,10]; // Vary proportional gain
  C = syslin('c',Kp+Ki/s+Kd*s);
```

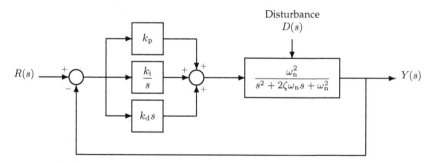

FIGURE 4.14 Second-order plant with proportional-integral-derivative control.

```
  Gclosed = (C*G)/.Gunity;
  y = csim('step',t,Gclosed);
  plot(t,y);
end
```

For Matlab, the equivalent code is:

```
s=tf('s');
zeta = 1; % Damping ratio
omegaN = 1; % Natural frequency
G = omegaN/(s^2+2*zeta*omegaN*s+omegaN^2); % System TF
t = 0:0.01:10;
Ki=1; % Integral gain
Kd=0; % Derivative gain
hold on;
for Kp = [2,5,10]; % Vary proportional gain
  C = syslin('c',Kp+Ki/s+Kd*s);
  Gclosed = feedback(C*G,1);
  step(Gclosed,t);
end
```

Figure 4.15 shows the time response for some different controller gain combinations. This example only looks at a PI controller (the derivative gain is set to zero) with a fixed integral gain and varying proportional gain. You can try other gain values and see the effect on the plant response. Try, for example, to find the best combination for (a) a plant that requires a fast response and can tolerate some overshoot and (b) a plant in which an overshoot in the response cannot be tolerated. We will look at similar cases of controller tuning at the end of this chapter.

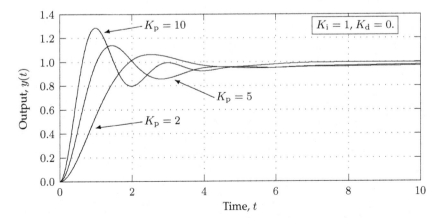

FIGURE 4.15 Time response plots for second-order system with proportional-integral control.

4.3.6 Summary: Controller features

Summarizing the findings from above regarding the different "control elements," we can list some characteristics of commonly used controller configurations.

- *Proportional control.* Proportional control alone cannot track the reference input; there will always be a steady-state error. However, proportional control can increase the speed of response. For larger gains we get a faster response and a lower steady-state error, but with an increasingly oscillatory response. Also, we cannot just increase the gain indefinitely, since the controller must interface with real-world actuators which have limitations both in terms of range (i.e., they saturate) and speed of response.

- *Proportional and derivative (PD) control.* A PD controller will not track the input, i.e., there will be some steady-state error. The derivative term provides damping and may therefore allow the use of higher proportional gain without giving excessive oscillations. PD control can be used when we want to avoid the destabilizing nature of integral control and when a steady-state error can be tolerated. It should be noted, however, that the derivative control element is sensitive to rapid changes in error signal. Derivative action will amplify any noise in the signal, and often a filter is used to take out high frequencies.

- *Proportional and integral (PI) control.* PI control gives a good speed of response through the proportional element while the integral action eliminates any steady-state error. PI control works well when the plant is predominantly first order. Higher order systems can be controlled with a PI controller if the control requirements are not too stringent

- *Proportional integral and derivative (PID) control.* A PID controller works well with second- or higher order systems. It is used when good transient response is desired. The derivative action gives a fast response without the output becoming oscillatory, and the integral term eliminates any steady-state error. However, as with PD control, the derivative action is susceptible to noise on the error signal.

4.4 DISTURBANCE REJECTION

Up to now, we have only looked at the response to a change in the reference input (i.e., the demand). However, equally important is the ability of a control system to reject a change in the disturbances, $D(s)$, to a system.

In order to simplify the analysis, we can assume that the disturbance is acting on the system input. This is a simplification because unwanted disturbances act directly on some component of the plant and may have a different influence than the ordinary system input, however this simplification allows us analyze factors such as controller disturbance rejection capability.

Obviously, it is desirable that the plant stay relatively insensitive to a disturbance input, i.e., that a change in the disturbance does not have a large effect on the output. The plant with a simplified disturbance input can be represented in block diagram form as in Figure 4.16.

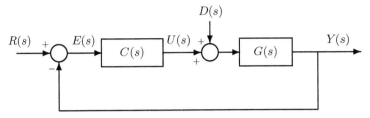

FIGURE 4.16 Feedback control system with disturbance input.

Previously, we used simple block diagram reduction techniques to find the total system transfer function. In this case, we have two inputs, and we will use a more fundamental technique which is useful to know when faced with more complex systems.

We would like to express the output $Y(s)$ as a function of the inputs $R(s)$ and $D(s)$. First, we can see that the input signal to the plant $G(s)$ is $D(s) + U(s)$, therefore the output $Y(s)$ can be written as

$$Y(s) = (D(s) + U(s))\, G(s).$$

But $U(s)$ is equal to the error signal $E(s)$ multiplied with the controller transfer function $C(s)$: $U(s) = E(s)C(s)$. Therefore, by substitution, we get

$$Y(s) = (D(s) + E(s)C(s))\, G(s).$$

Going further backwards in the loop, we have that $E(s) = R(s) - Y(s)$, hence the above equation becomes:

$$Y(s) = (D(s) + [R(s) - Y(s)]\, C(s))\, G(s).$$

This equation gives the output $Y(s)$ as a function of the two inputs $R(s)$ and $D(s)$ with given controller and plant transfer functions, $C(s)$ and $G(s)$. Solving for $Y(s)$ and rearranging, we have

$$Y(s) = \frac{G(s)}{1 + C(s)G(s)} D(s) + \frac{C(s)G(s)}{1 + C(s)G(s)} R(s).$$

Note that if $D(s) = 0$, this is the same as we get using the block diagram reduction rules in the examples above.

Now, let us look at the response of the system to changes in the reference input or the disturbance. Assume that we have a second-order plant with transfer function $G(s) = \dfrac{K\omega_n^2}{s^2 + 2\zeta\omega_n s + \omega_n^2}$, where K is the plant gain, and a proportional controller with gain K_p. Since we can use the principle of superposition (see Chapter 2), we can look at the inputs individually.

The closed-loop transfer function when $D(s)$ to zero is

$$\frac{Y(s)}{R(s)} = \frac{C(s)G(s)}{1 + C(s)G(s)} = \frac{K_p K \omega_n^2}{s^2 + 2\zeta \omega_n s + \omega_n^2 + K_p K \omega_n^2}.$$

Using the final value theorem, we find the steady-state response to a unit step input:

$$\lim_{t \to \infty} y(t) = \lim_{s \to 0} sY(s) = s \cdot \frac{K_p K \omega_n^2}{s^2 + 2\zeta \omega_n s + \omega_n^2 + K_p K \omega_n^2} \frac{1}{s} = \frac{K_p K}{1 + K_p K}.$$

This is the same as we found for the proportional controller above (see page 77), only in that case the two gain variables were merged into one single variable. As above, we see that there will always be a steady-state error when using proportional control alone.

In order to look at the response to a step change in the disturbance, we set $R(s) = 0$. The closed-loop transfer function from disturbance $D(s)$ to output $Y(s)$ is

$$\frac{Y(s)}{D(s)} = \frac{G(s)}{1 + C(s)G(s)} = \frac{K \omega_n^2}{s^2 + 2\zeta \omega_n s + \omega_n^2 + K_p K \omega_n^2}.$$

The steady-state response to a step change in $D(s)$ is:

$$\lim_{t \to \infty} y(t) = \lim_{s \to 0} sY(s) = s \cdot \frac{K \omega_n^2}{s^2 + 2\zeta \omega_n s + \omega_n^2 + K_p K \omega_n^2} \frac{1}{s} = \frac{K}{1 + K_p K}.$$

Although this is not equal to zero, and the disturbance therefore is not rejected completely, the error introduced by the disturbance can be reduced by increasing the controller gain K_p.

4.5 EXAMPLES OF FEEDBACK CONTROL SYSTEMS

In this section, we will look at two examples of feedback control systems and their analysis.

4.5.1 Electric pump drive speed control

In the last chapter, we derived a model for an electric pump drive (see Figure 3.21). Suppose we would like to control the pump with a PID controller. We can build the controller around the pump drive block diagram model, as shown in Figure 4.17. As can be seen, the controller receives the error between the reference input $R(s)$ and the actual output speed $s\theta(s)$, and adjusts the input voltage by means of proportional, integral, and derivative gain terms to try and obtain the desired motor speed.

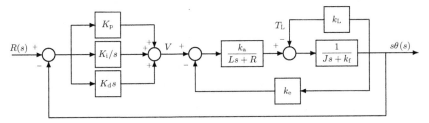

FIGURE 4.17 Electric pump drive with PID control.

We can now adapt the simulation code developed before (see page 64) to include the controller and feedback loop. For Scilab, the following code will be suitable:

```
// Helping variables:
  s=%s;
  Gunity=syslin('c',1,1); // Needed for unity feedback
      in Scilab
// Motor coefficients:
  L = 0.4;
  R = 1;
  ka = 0.5;
  kL = 1;
  J = 0.2;
  kf = 0.2;
  ke = 0.5;
// Motor transfer function:
  Gmotor = syslin('c',ka/(J*L*s^2 + (J*R+kf*L+kL*L)*s +
      (kf*R+kL*R+ka*ke)));
// Controller coefficients and transfer function:
  Kp=1;
  Ki=0;
  Kd=0;
  C = syslin('c',Kp+Ki/s+Kd*s);
// Total (feedback) transfer function:
  G = (C*Gmotor)/.Gunity;
// Simulate response for a step input of 10 and plot output:
  t = 0:0.01:3;
  y = 10*csim('step',t,G);
  plot(t,y);
```

Equivalent Matlab code is:

```
% Helping variables:
  s=tf('s');
% Motor coefficients:
  L = 0.4;
  R = 1;
  ka = 0.5;
  kL = 1;
  J = 0.2;
  kf = 0.2;
```

```
ke = 0.5;
% Motor transfer function:
Gmotor = ka/(J*L*s^2 + (J*R+kf*L+kL*L)*s + (kf*R+kL*R+ka*ke));
% Controller coefficients and transfer function:
Kp=1;
Ki=0;
Kd=0;
C = Kp+Ki/s+Kd*s;
% Total (feedback) transfer function:
G = feedback(C*Gmotor,1);
% Simulate response for a step input of 10 and plot output:
t = 0:0.01:3;
step(G,t);
```

Now we can implement different PID controllers by inserting values for K_p, K_i, and K_d in the script and running the model. For a purely proportional controller, let us try values of K_p of 1, 10, 20, and 30 (while letting $K_i = K_d = 0$).

Figure 4.18 shows the results, and we see that increasing K_p gives a faster initial response and the steady-state output gets closer to the reference value of 10, but it also increases the magnitude of the overshoot and oscillations. Let us settle for a value of $K_p = 20$. We can never get a zero steady-state error with a proportional controller alone, so let us try some integral control as well.

Figure 4.19 shows the output for the PI controller. For low integral gain ($K_i = 5$), the steady-state error is only slowly eliminated but the influence on system dynamics is minimal. Increasing K_i gives a more rapid elimination of the steady-state error but increases the oscillations, in the worst case potentially making the system unstable. Let us use a value of 30 for K_i and try some differential feedback as well.

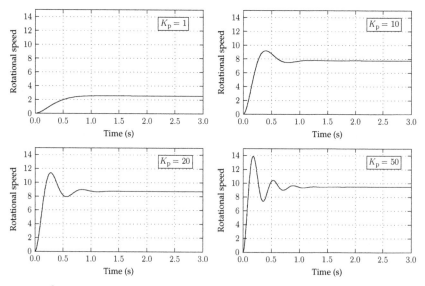

FIGURE 4.18 Electric pump drive with P-only control.

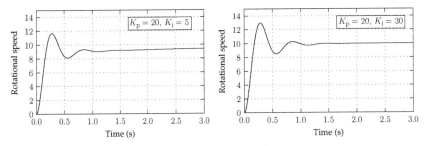

FIGURE 4.19 Electric pump drive with PI control.

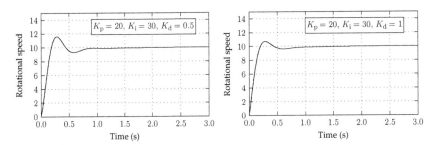

FIGURE 4.20 Electric pump drive with PID control.

It can be seen from Figure 4.20 that adding differential control action adds damping to the system, which reduces oscillations, and also speeds up the response.

This example has shown the influence of the different controller elements, however, we should note that this is an ideal simulation model; in practice, there may be limits to our freedom to tune the controller. For example, we probably cannot get the actuator to produce whatever output we want, there may be a maximum voltage V available to excite the motor. In such a case, it does not matter what the controller demands; we won't get more than the maximum anyway. (The actuator saturates; see Section 2.4.2.)

4.5.2 Ship autopilot control

A free body diagram of a ship with autopilot heading control is shown in Figure 4.21. (Example adapted from Ref. [1], with permission.) The relevant system variables are

- the desired heading, $\psi_d(t)$,
- the actual heading, $\psi_a(t)$, and
- the rudder angle, $\delta(t)$.

In Figure 4.21, X_0, Y_0 is the coordinate system, where X_0 is aligned to the north. All angles are in radians and measured with respect to X_0, with the exception of the rudder angle, δ, as shown.

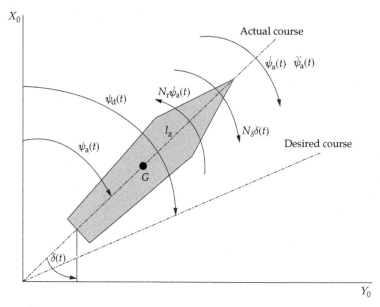

FIGURE 4.21 Free body diagram of ship hull for heading control.

The equation of motion for the yaw axis, taking into account the resultant moment of force acting on the hull from the rudder and from the hull resistance to the yaw motion, is

$$\sum M_G = I_Z \ddot{\psi}_a(t)$$

$$N_\delta \delta(t) - N_r \dot{\psi}_a(t) = I_Z \ddot{\psi}_a(t),$$

where N_δ and N_r are hydrodynamic coefficients associated with the rudder and the hull, respectively.

Taking Laplace transforms:

$$N_\delta \delta(s) = (I_Z s^2 + N_r s) \psi_a(s).$$

Hence, the hull transfer function becomes

$$\frac{\psi_a(s)}{\delta(s)} = \frac{N_\delta}{s(I_Z s + N_r)}.$$

Building a feedback control system for heading control we get a block diagram as shown in Figure 4.22.

A gyro-compass provides a measured heading value, ψ_m, which is equal to the actual heading, ψ_a, multiplied with the gyro-compass feedback gain H_1:

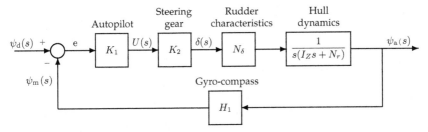

FIGURE 4.22 Block diagram of ship autopilot control system.

$$\psi_m(s) = H_1 \psi_a(s).$$

The controller is assumed to provide proportional control only, with a gain K_1 on the error signal, i.e., the difference between the desired (ψ_d) and the measured (ψ_m) headings. The input signal to the steering gear is, therefore,

$$U(s) = K_1(\psi_d(s) - \psi_m(s)).$$

For this example, we shall assume that the dynamics associated with the steering gear can be ignored, hence the rudder angle $\delta(s)$ is simply proportional to the steering gear signal by the factor K_2.

We can now work out the closed-loop transfer function for the system:

$$\frac{\psi_a(s)}{\psi_d(s)} = \frac{\frac{K_1 K_2 N_\delta}{s(I_Z s + N_r)}}{1 + \frac{K_1 K_2 N_\delta H_1}{s(I_Z s + N_r)}},$$

which can be simplified to

$$\frac{\psi_a(s)}{\psi_d(s)} = \frac{1/H_1}{\left(\frac{I_Z}{K_1 K_2 N_\delta H_1}\right)s^2 + \left(\frac{N_r}{K_1 K_2 N_\delta H_1}\right)s + 1}.$$

For a specific hull, the control problem is to determine the autopilot setting (K_1) to provide a satisfactory transient response. In this case, this will be when the damping ratio has a value of 0.5. Also to be determined are the rise time, settling time, and percentage overshoot in the response.

The ship to be controlled is a cargo vessel of length 161 m with a MARINER-type hull with a total displacement of 17,000 ton. The system coefficients are:

- $K_2 = 1.0\,\text{rad/V}$
- $N_r = 2 \times 10^9\,\text{N ms/rad}$
- $H_1 = 1.0\,\text{V/rad}$
- $N_\delta = 80 \times 10^6\,\text{N m/rad}$
- $I_Z = 20 \times 10^9\,\text{kg m}^2$

Inserting the values into the system equation gives

$$\frac{\psi_a(s)}{\psi_d(s)} = \frac{1}{\left(\frac{20\times 10^9}{K_1\times 80\times 10^6}\right)s^2 + \left(\frac{2\times 10^9}{K_1\times 80\times 10^6}\right)s + 1},$$

which simplifies to

$$\frac{\psi_a(s)}{\psi_d(s)} = \frac{\frac{K_1}{250}}{s^2 + \frac{1}{10}s + \frac{K_1}{250}}.$$

We can compare this to the standard form (see page 42) to work out the required value for K_1. Equating the first-order coefficients we have

$$2\zeta\omega_n = \frac{1}{10}.$$

Knowing that $\zeta = 0.5$ (which is the requirement), we get $\omega_n = 1/10$. Equating the constant terms in the characteristic equation, we can find a value of K_1 to produce the required damping ratio:

$$\frac{K_1}{250} = \omega_n^2,$$
$$\frac{K_1}{250} = \frac{1}{100},$$
$$K_1 = 2.5.$$

This system can be simulated using the software similarly as above. Figure 4.23 shows the response of the vessel to a unit step input in the desired heading $\psi_d(s)$ with $K_1 = 2.5$. The performance parameters are:
- Rise time to 95%: 23 s.
- Percentage overshoot: 16.3%.
- Settling time (to ±2%): 81 s.

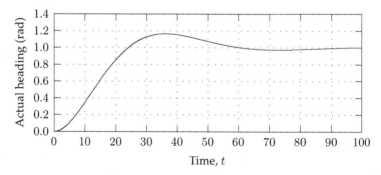

FIGURE 4.23 Unit step response of the ship autopilot system.

By adapting the simulation code from the previous example to this case, one could also investigate the effect of varying K_1 over a wider range of values and, e.g., what would happen if we used a more realistic representation of the steering gear dynamics. (Try, e.g., to replace the simple gain term K_2 with a first-order transfer function with a time constant of a few seconds.)

QUESTIONS

4.1 For a first-order system with a transfer function $\dfrac{1}{s+2}$, calculate the time constant. Where are the poles and zeros? How does the time constant change if we apply a unity gain feedback loop to the system? What happens to the poles and zeros? How does the system change if you use a compensator *in the feedback loop* with a gain of 10? What happens to the responsiveness of the system?

4.2 A simplified version of the d.c. motor model from the pump drive example above is shown in the figure. The input is voltage, V, and the output is rotational speed, N.

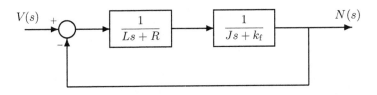

(a) What is the transfer function for the d.c. motor?

(b) If the system is controlled with a proportional and integral controller with gains K_p and K_i, what is the total system transfer function?

4.3 A marine engine has a transfer function $\dfrac{1}{s(s+5)}$ at 100 rpm, where the input is fuel flow rate and the output is rotational speed. We control the fuel valve with an electric motor which has a transfer function $\dfrac{1}{s+1}$, where the input is voltage and the output is fuel flow rate. We implement a PI controller for the motor with gains of K_p and K_i, respectively. A tachometer in the feedback loop measures the rotational speed and feeds back with a gain of 5.

(a) Draw the system block diagram and derive the closed-loop transfer function (in standard form).

(b) At the lower speed of 70 rpm the engine transfer function is $\dfrac{1}{s(s+2)}$. Calculate the total system transfer function if the engine runs at 70 rpm.

4.4 It was shown above (Section 4.3.4) that the velocity feedback controller, shown below, tracks the reference input (zero steady-state error). For a plant $G(s) = \dfrac{1}{(s+a)(s+b)}$, work out the transfer function between the disturbance $D(s)$ and the output $Y(s)$ (assuming $R(s) = 0$) and check whether this controller will reject a unit step input in the disturbance $D(s)$.

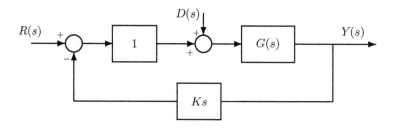

REFERENCE

[1] R.S. Burns, Advanced Control Engineering, Butterworth Heinemann, Oxford, 2001.

Chapter Five

Closed-Loop Stability

CHAPTER POINTS

- Routh-Hurwitz stability criterion
- Root locus plots
- Examples of sketching root locus plots
- Root locus in Scilab and Matlab

5.1 ROUTH-HURWITZ STABILITY CRITERION

We have seen that a system representation in the s domain will have a characteristic equation (the denominator of the system transfer function is set to zero) of the form

$$a_0 s^n + a_1 s^{n-1} + a_2 s^{n-2} + \cdots + a_{n-1} s + a_n = 0.$$

In Section 3.3.1 (page 48), we saw that for a system to be stable, all the poles of the system, i.e., the roots of the characteristic equation, must have negative (or zero) real parts.

Stability is vital in any control system, and therefore determining whether a system is stable or unstable is of key importance to the control engineer. In order to determine whether a system is stable, we either calculate all the roots (poles) and check their value, or use a stability criterion. A stability criterion is a method of determining whether a system is stable without actually calculating the roots of the characteristic equation.

THE ROUTH-HURWITZ STABILITY CRITERION

The Routh-Hurwitz stability criterion states that for a system having a characteristic equation

$$a_0 s^n + a_1 s^{n-1} + a_2 s^{n-2} + \cdots + a_{n-1} s + a_n = 0$$

Marine Systems Identification, Modeling, and Control. http://dx.doi.org/10.1016/B978-0-08-099996-8.00005-7

to be asymptotically stable, all the principal minors[1] of the matrix

$$
H_n =
\begin{pmatrix}
a_1 & a_3 & a_5 & \cdots & \cdots & 0 \\
a_0 & a_2 & a_4 & \cdots & \cdots & 0 \\
0 & a_1 & a_3 & a_5 & \cdots & 0 \\
0 & a_0 & a_2 & a_4 & \cdots & 0 \\
0 & 0 & a_1 & a_3 & \cdots & 0 \\
\vdots & \vdots & \vdots & \vdots & & \vdots \\
0 & 0 & \cdots & \cdots & \cdots &
\end{pmatrix}
$$

must be *positive, nonzero*. (The matrix H_n is known as the *Hurwitz matrix*.) For example, consider the characteristic equation

$$2.5s^4 + 1.3s^3 + 0.02s^2 - s + 5 = 0.$$

This is a fourth-order system ($n = 4$), therefore H_n is a 4×4 matrix. Compare the characteristic equation and the matrix which needs to be populated:

$$a_0 s^4 + a_1 s^3 + a_2 s^2 + a_3 s + a_4 = 0$$

$$
H_4 =
\begin{pmatrix}
a_1 & a_3 & 0 & 0 \\
a_0 & a_2 & a_4 & 0 \\
0 & a_1 & a_3 & 0 \\
0 & a_0 & a_2 & a_4
\end{pmatrix}
$$

Substituting, we have

$$
H_4 =
\begin{pmatrix}
1.3 & -1 & 0 & 0 \\
2.5 & 0.02 & 5 & 0 \\
0 & 1.3 & -1 & 0 \\
0 & 2.5 & 0.02 & 5
\end{pmatrix}
$$

For stability, $\Delta_1, \Delta_2, \Delta_3, \Delta_4 > 0$

$$\Delta_1 = 1.3 > 0 \ \checkmark$$

$$\Delta_2 = \begin{vmatrix} 1.3 & -1 \\ 2.5 & 0.02 \end{vmatrix} = 2.526 > 0 \ \checkmark$$

$$\Delta_3 = \begin{vmatrix} 1.3 & -1 & 0 \\ 2.5 & 0.02 & 5 \\ 0 & 1.3 & -1 \end{vmatrix} = -10.976 < 0$$

---

[1] See Appendix B.3.

The two first determinants are positive, however, the third is not. Thus, the system is not stable. (We don't have to calculate the fourth determinant.)

One advantage of this method is that the analysis can be done in terms of system parameters, and stability conditions can therefore be derived. Consider the mass-spring-damper system (see page 40), which has a characteristic equation

$$ms^2 + Cs + k = 0.$$

For a second-order system, we only get a 2×2 matrix, and the analysis is straightforward. We have $a_0 = m$, $a_1 = C$, and $a_2 = k$. Note that there is no a_3, so we set that entry to zero in the Hurwitz matrix.

$$H_2 = \begin{pmatrix} a_1 & a_3 \\ a_0 & a_2 \end{pmatrix} = \begin{pmatrix} C & 0 \\ m & k \end{pmatrix}$$

$$\Delta_1 = C > 0$$

$$\Delta_2 = \begin{vmatrix} C & 0 \\ m & k \end{vmatrix} = Ck > 0.$$

In other words, for such systems to be stable, $C > 0$ and $Ck > 0$. (But C and k are the damper and spring constants, which will always be positive, therefore the spring-mass-damper system in its simplest form will never be unstable.)

LIÉNARD-CHIPART RULE
It can still be a pain to calculate these determinants; however, Liénard and Chipart showed that if all the coefficients in the characteristic equation are positive ($a_i > 0$ for $i = 1, \ldots, n$) then the system is asymptotically stable if either

$$\Delta_1, \Delta_3, \Delta_5, \ldots > 0$$

or

$$\Delta_2, \Delta_4, \Delta_6, \ldots > 0.$$

Thus only half the number of determinants needs to be evaluated.
For example, with a characteristic equation and corresponding Hurwitz matrix of:

$$3s^3 + 3s^2 + 9s + 6 = 0$$

$$H_3 = \begin{pmatrix} 3 & 6 & 0 \\ 3 & 9 & 0 \\ 0 & 3 & 6 \end{pmatrix}$$

Since all coefficients are positive, $a_i > 0$, we only need to consider either Δ_1 and Δ_3, or Δ_2. Looking at Δ_1 and Δ_3, we have

$$\Delta_1 = 3 > 0 \;\checkmark$$

$$\Delta_3 = \begin{vmatrix} 3 & 6 & 0 \\ 3 & 9 & 0 \\ 0 & 3 & 6 \end{vmatrix} = 54 > 0 \;\checkmark$$

$\Rightarrow$ system is stable.

Or, we can save ourselves some work and only look at Δ_2:

$$\Delta_2 = \begin{vmatrix} 3 & 6 \\ 3 & 9 \end{vmatrix} = 9 > 0 \;\checkmark$$

$\Rightarrow$ system is stable.

Another example, suppose we have a system

$$\frac{Y(s)}{U(s)} = \frac{K}{s^3 + 3s^2 + 2s + K}.$$

The characteristic equation is

$$s^3 + 3s^2 + 2s + K = 0$$

and the corresponding 3×3 Hurwitz matrix is

$$H_3 = \begin{pmatrix} 3 & K & 0 \\ 1 & 2 & 0 \\ 0 & 3 & K \end{pmatrix}.$$

We can apply the Liénard and Chipart rule because K is a gain and always positive nonzero.[2] The easiest option is to consider Δ_2 only:

$$\Delta_2 = \begin{vmatrix} 3 & K \\ 1 & 2 \end{vmatrix} = 6 - K > 0.$$

Hence, this system is stable for $0 < K < 6$. We can test this quickly by plotting the system step response for K-values of, say, 5, 6, and 7, as shown in Figure 5.1.

The simulation code to produce this plot is shown below for Scilab (left) and Matlab (right).

---

[2] It is possible to have a negative gain but we will not encounter such systems here.

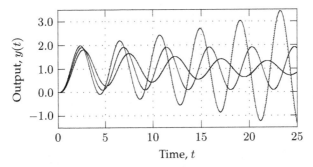

FIGURE 5.1 Effect of varying K on system dynamic response.

```
s=%s;
t=0:0.01:25;

for K=[5,6,7]
  G=syslin('c',K/(s^3+3*s^2+2*
    s+K));
  y=csim('step',t,G);
  plot(t,y);
end
```

```
s=tf('s');
t=0:0.01:25;

hold on;
for K=[5,6,7]
  G=K/(s^3+3*s^2+2*s+K);
  step(G,t);
end
```

Therefore, if K is a controller gain, we need to take care when tuning the controller since increasing the gain too much will drive the system to instability.

5.2 ROOT LOCUS

Following from the theory and examples from the last chapter, root locus is a technique to study what happens with a system if we implement a feedback control loop around it and then gradually increase the controller gain.

EXAMPLE: FIRST-ORDER SYSTEM

We saw in Section 4.2.2 that for a plant with an open-loop transfer function $\frac{K}{\tau s + 1}$, the closed-loop transfer function is

$$\frac{K}{\tau s + K + 1},$$

which has a pole at $s = -(K/\tau + 1/\tau)$. For very low gain values ($K \ll 1$), the closed-loop pole is close to that of the open-loop system. As the gain increases, the pole becomes more negative, which corresponds to shorter time constant (i.e., a faster response). We can plot this to see the effect of the gain on pole placement, as shown in Figure 5.2.

The simulation code to produce this plot is shown below for Scilab (left) and Matlab (right).

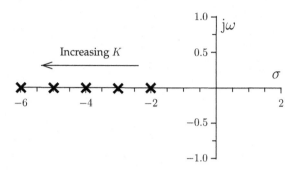

FIGURE 5.2 System pole placement as a function of gain K.

```
s = %s;                          s = tf('s');
tau = 1;                         tau = 1;
for K = 1:5;                     hold on;
  G = syslin('c',K/(s+K/tau+1/   for K = 1:5;
      tau));                       G = K/(s+K/tau+1/tau);
  plzr(G); // Create pole-zero     pzplot(G); % Create pole-
      map                              zero map
end                              end
```

In this example, we used only five gain values; however, we could populate this plot with more pole values for different gains. (For example, try using K=0.1:0.1:10 in the for loop.) But what if we were to plot the pole placement for *all* values of K from zero to infinity, with infinitely small increments? In that case, we would get a graph which would show the system poles for any imaginable gain value. That would be a root locus plot.

For the current system, this is straightforward: we get a graph starting at -1 (which is the closed system pole for $K = 0$), and shooting off to minus infinity on the real axis. Figure 5.3 shows the actual root locus plot for this system. Whatever the value of K, the pole will never have an imaginary part. Of course that is not always the case, and we will see considerably more complicated root locus plots later.

EXAMPLE: SECOND-ORDER SYSTEM

Consider a plant with open-loop transfer function $\dfrac{0.75}{s(s+1)}$ which we want to control with a proportional controller, K, as shown in Figure 5.4.

The closed-loop transfer function is

$$\frac{Y(s)}{U(s)} = \frac{K\frac{0.75}{s(s+1)}}{1 + K\frac{0.75}{s(s+1)}} = \frac{0.75K}{s^2 + s + 0.75K}.$$

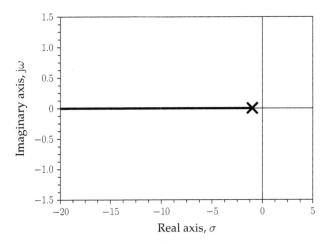

FIGURE 5.3 Root locus plot for $G(s) = \dfrac{K}{1 + s\tau}$.

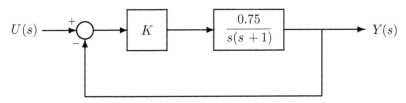

FIGURE 5.4 Closed-loop system with proportional control.

The closed-loop poles are found by solving the characteristic equation:

$$s^2 + s + 0.75K = 0 \Rightarrow s = \frac{-1 \pm \sqrt{1 - 3K}}{2}.$$

We see that if $(1 - 3K) < 0$, the roots will be complex. So we have

$$s = \begin{cases} \dfrac{-1 \pm \sqrt{1 - 3K}}{2} & \text{if } K \le \dfrac{1}{3} \\[2ex] \dfrac{-1 \pm j\sqrt{3K - 1}}{2} & \text{if } K > \dfrac{1}{3} \ \ (\text{remember that } j = \sqrt{-1}). \end{cases}$$

If $K = 0$, the poles are at 0 and -1. As K increases, the pole at zero becomes more negative and the pole at -1 becomes more positive (while $K < \frac{1}{3}$). In other words, the two poles move toward each other.

At $K = \frac{1}{3}$, both poles are at -0.5 (since the square root term becomes zero). For $K > \frac{1}{3}$, the poles become complex, with constant real value -0.5 and opposite imaginary values increasing in magnitude with increasing K. These values are known as a complex conjugate pair. Figure 5.5 shows the root locus plot for the system.

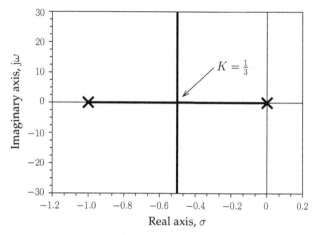

FIGURE 5.5 Root locus plot for $G(s) = \dfrac{0.75}{s(s+1)}$.

From the above analysis we can see that for $K > \frac{1}{3}$, the system will become oscillatory and the frequency will increase with increasing K. The oscillations will be damped since the real part of s (i.e., σ) is negative. If the root locus were to go beyond the y-axis into the right-hand plane (i.e., $\sigma > 0$), we would know that the system would be unstable for that value of K.

5.2.1 Characteristics of root locus plots

As shown above, by looking at the characteristic equation of the closed-loop transfer function, we can see whether the system will become oscillatory, or even unstable, for increasing values of controller gain. But do we have to work out the closed-loop transfer function every time?

Consider a standard unity feedback system as in Figure 5.6. The closed-loop transfer function is

$$\frac{Y(s)}{U(s)} = \frac{KG(s)}{1 + KG(s)}$$

It can be seen that we only need to solve $1 + KG(s) = 0$ to plot the closed-loop poles, and this is a function of the open-loop transfer function $KG(s)$. Therefore, the root locus analysis provides a convenient method of displaying the location of the closed-loop poles using the open-loop transfer function, $KG(s)$.

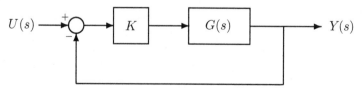

FIGURE 5.6 Unity feedback system.

BACKGROUND AND MATHEMATICAL DERIVATION

The characteristic equation, $1 + KG(s) = 0$, can be written as

$$1 + K\frac{(s - z_1)(s - z_2)\ldots(s - z_u)}{(s - p_1)(s - p_2)\ldots(s - p_v)} = 0,$$

where z_i, $i = 1, 2, \ldots, u$ and p_i, $i = 1, 2, \ldots, v$ are the zeros and poles of the open-loop transfer function, $G(s)$. If $(s - z_1)(s - z_2)\ldots(s - z_u) = Z(s)$ and $(s - p_1)(s - p_2)\ldots(s - p_v) = P(s)$, the closed-loop characteristic equation can be written as

$$P(s) + KZ(s) = 0.$$

Remember that the solutions of this equation determine the poles of the closed-loop system. For $K = 0$, the roots of this equation are the same as the poles of the open-loop transfer function ($P(s)$), and for very large values of K, the roots tend to the zeros of the open-loop transfer function ($Z(s)$). Thus, as K is increased from zero to infinity, the roots of the closed-loop characteristic equation start at the poles of the open-loop transfer function and terminate at the zeros of the open-loop transfer function. Hence, from this we know where the root loci start and end.

Since we are working with complex numbers, we know that in the equation $KG(s) = -1$, both the argument and the magnitude must be the same on both sides.[3] The right-hand side, -1, has modulus 1 and argument $-\pi$ (or $-180°$). Therefore, the angle of the complex number $KG(s)$ must be equal to $\pi + k \cdot 2\pi$, where k is an integer. K has argument 0,[4] so this implies that

$$\arg(G(s)) = \pi + 2\pi k.$$

We know from above that $G(s) = \dfrac{(s - z_1)(s - z_2)\ldots(s - z_u)}{(s - p_1)(s - p_2)\ldots(s - p_v)}$, and using the calculation rules (see Appendix B.1) we get

$$\sum_{i=1}^{u} \arg(s - z_i) - \sum_{i=1}^{v} \arg(s - p_i) = (1 + 2k)\pi.$$

These relationships are known as the angle criteria. In addition, the magnitude of number $G(s)$ must equal $\left|\dfrac{1}{K}\right|$. Thus,

$$|K| = \frac{\prod_{i=1}^{v} |(s - p_i)|}{\prod_{i=1}^{u} |(s - z_i)|}.$$

This is known as the magnitude criteria.

---

[3] See Appendix B.1 (page 151) for more information on complex numbers and the background for this derivation.

[4] This applies for $K \geq 0$ only, but we will not come across negative gains here.

ROOT LOCUS CONSTRUCTION RULES

Using the above criteria, it is possible to plot the paths (loci) taken by the roots of the closed-loop characteristic equation as K is varied from zero to infinity. To get an accurate root locus plot, computational tools are usually needed. However, there are a number of simple rules and guidelines which can be used to construct approximate root locus plots. These are:

1. The number of branches of the root locus plot, i.e., the number of loci, is equal to the number of poles of the open-loop transfer function (in other words, the degree of the characteristic equation).

2. For any imaginary solutions of the characteristic equation, the complex roots occur in conjugate pairs. Thus, the loci are symmetric about the real axis.

3. The loci start at the poles of the open-loop transfer function (for $K = 0$). These points are shown by a $\times$ on the plot.

4. The loci end at the zeros of the open-loop transfer function (for $K = \infty$). These points are shown by a $\odot$ on the plot.

5. If $G(s)$ has r more poles than zeros ($v - u = r$), then r loci will go to infinity, asymptotic to r straight lines making angles φ with the real axis of

$$\varphi = (2k + 1)\frac{\pi}{r},$$

where $k = 0, 1, 2, \ldots, (r - 1)$.

6. The asymptotes intersect the real axis at the point a_0, where

$$a_0 = \frac{\left(\sum_{i=1}^{v} p_i - \sum_{i=1}^{u} z_i\right)}{r}$$

and p_i and z_i are the poles and zeros of the open-loop transfer function.

7. Going from $-\infty$ to ∞ on the real axis, if at a given point the sum of real poles and zeros to the right is odd, then a locus exist at this point on the real axis.

8. A breakaway point is a location on the real axis where the root locus branches either arrive or depart from the real axis (see Figure 5.7).

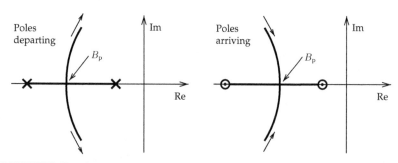

FIGURE 5.7 Root locus breakaway points.

These points are given by

$$\sum_{i=1}^{v} \frac{1}{(B_p - p_i)} = \sum_{i=1}^{u} \frac{1}{(B_p - z_i)}.$$

In general, this equation can only be solved by hand for simple system transfer functions; with a more complex $G(s)$, an iterative method such as Newton-Raphson can be used.

9. The intersection of the root loci with the imaginary axis (i.e., the value of K at the limit of stability) can be found using the Routh-Hurwitz stability criterion.

10. The departure angle of the root locus from a complex pole (see Figure 5.8) is given by

$$\theta_d = \pi - \sum_{i}^{v} \angle p_i + \sum_{i}^{u} \angle z_i,$$

where $\angle p_i$ and $\angle z_i$ are the angles from the pole in question to the other open-loop poles and zeros.

(For the system in the illustration, the departure angle is $\theta_d = \pi - \theta_{p1} - \theta_{p2} + \theta_{z1}$. A similar analysis must be done for the second complex pole.) Similarly, the angle of arrival of the root locus at a complex zero is given by

$$\theta_a = \pi + \sum_{i}^{v} \angle p_i - \sum_{i}^{u} \angle z_i.$$

These are the 10 construction rules for sketching root locus plots. Don't worry if you could not follow the derivation of the rules at the moment, it will become clearer when we look at some examples in the following sections.

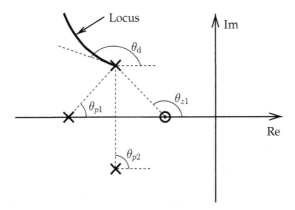

FIGURE 5.8 Root locus departure angle from a complex pole.

5.2.2 Examples of sketching root locus plots

EXAMPLE 1

Draw the root locus for the system $KG(s) = \dfrac{K(s+3)}{s(s+2)(s+5)}$. ($KG(s)$ is the open-loop transfer function.)

We apply the rules:

1. There are three open-loop poles, therefore three loci branches.
2. Any imaginary loci are symmetrical about the real axis.
3. The loci start at the open-loop poles: $0, -2, -5$.
4. The loci terminate at the open-loop zeros: -3. Since there are three branches, one will terminate at -3 while the other two will go to infinity. We can now start a sketch (see Figure 5.9).
5. Next, we find the asymptotic angles for the infinite loci. We have the number of zeros, $u = 1$, the number of poles, $v = 3$, hence $r = v - u = 3 - 1 = 2$, and the angles of the asymptotes are

$$\varphi = (2k+1)\frac{\pi}{r} \quad \text{for } k = 0, 1 :$$
$$k = 0 \Rightarrow \varphi = \frac{\pi}{2},$$
$$k = 1 \Rightarrow \varphi = \frac{3\pi}{2}.$$

In other words, the two infinite branches go off at angles $90°$ and $270°$ with the real axis.

6. The asymptotes intersect the real axis at the point

$$a_0 = \frac{\sum_{i=1}^{v} p_i - \sum_{i=1}^{u} z_i}{r} = \frac{((0 - 2 - 5) - (-3))}{2} = -2.$$

Hence, we can add helping lines for the asymptotes, which in this case are vertical and intersecting -2 on the real axis.

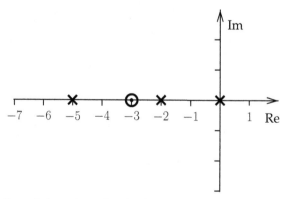

FIGURE 5.9 Example 1, root locus first sketch.

7. To find out where the loci lie on the real axis, imagine that we walk along the real axis and count the number of obstacles in front of us (only counting those that lie *on* the real axis). Between -7 and -5, there are four (three poles and one zero); this is an even number and this section is therefore not part of a locus. Between -5 and -3 there are three (an odd number), i.e., this section is part of a locus and we mark this in the figure. Similarly, between -2 and 0 we have one pole in front of us, therefore this section is also part of the locus.

The updated sketch with the asymptotes from Rules 5 and 6, as well as the parts of the loci that lie on the real axis, is shown in Figure 5.10.

8. From the figure, we see that the pole starting at -5 will terminate in the zero at -3, whereas the other two poles will move toward each other and then break away from the real axis and go to infinity along the asymptotes. The breakaway point is not necessarily the midpoint between the two open-loop poles, so we need to calculate it. We have

$$\sum_{i=1}^{v} \frac{1}{(B_p - p_i)} = \sum_{i=1}^{u} \frac{1}{(B_p - z_i)}$$

$$\frac{1}{B_p} + \frac{1}{B_p + 2} + \frac{1}{B_p + 5} = \frac{1}{B_p + 3}.$$

Rearranging and simplifying this, we get

$$B_p^3 + 8B_p^2 + 21B_p + 15 = 0.$$

We can't solve this by hand, but we know that B_p lies somewhere between 0 and -2, so we can try some values and possibly get an approximate solution. Trying $B_p = -1$, the left-hand side comes out at 1. $B_p = -1.1$ gives 0.249, whereas $B_p = -1.15$ gives 0.09. Thus, $B_p \approx -1.15$.

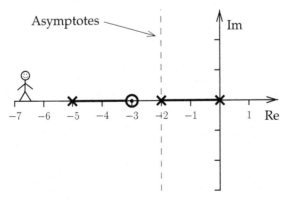

FIGURE 5.10 Example 1, root locus second sketch.

9. The system is stable for all $K \geq 0$ (we can prove this with the Routh-Hurwitz criterion), hence there will be no crossing of the loci into the positive real half plane.
10. There are no complex open-loop poles or zeros, therefore we don't have to calculate any departure or arrival angles.

The loci will break away from the real axis at a right (90°) angle, but we don't know exactly how the loci will go from the breakaway point toward the asymptotes and infinity. However, since we are only making a sketch, a graphical estimate of this is sufficient. Important is that the poles do not cross into the positive real half plane on the way (which would give instability for certain values of K), but if that were the case we would have found out in step 9 above. The final root locus sketch is shown in Figure 5.11.

EXAMPLE 2

Consider the transfer function $G(s) = \dfrac{100}{s(s^2 + 15s + 90)}$. By applying the 10 rules we can sketch the root locus:

1. There are three open-loop poles, therefore three loci branches.
2. Loci are symmetrical about the real axis.
3. The loci start at the open-loop poles: $0, -7.5 \pm j5.8$.
4. Since there are no zeros, all three branches will go to infinity.
5. We have no zeros ($u = 0$), three poles ($v = 3$), hence $r = v - u = 3$. The asymptotic angles are:

$$\varphi = (2k + 1)\frac{\pi}{r} \quad \text{for } k = 0, 1, 2:$$
$$k = 0 \Rightarrow \varphi = \frac{\pi}{3},$$
$$k = 1 \Rightarrow \varphi = \pi,$$
$$k = 2 \Rightarrow \varphi = \frac{5\pi}{3}.$$

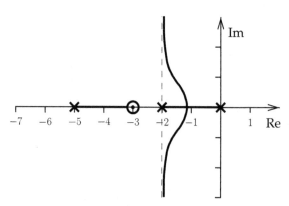

FIGURE 5.11 Example 1, final root locus sketch.

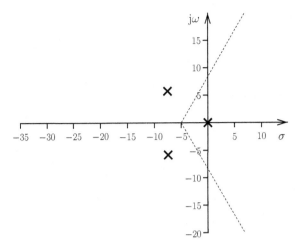

FIGURE 5.12 Example 2, root locus first sketch.

6. The asymptotes intersect the real axis at the point

$$a_0 = \frac{\sum_{i=1}^{v} p_i - \sum_{i=1}^{u} z_i}{r} = (0 + (-7.5 - j5.8) + (-7.5 + j5.8))/3 = -5.$$

The sketch so far, with loci starting points and asymptotes, is shown in Figure 5.12. (Note that the asymptote with angle π follows the real axis to $-\infty$.)

7. By inspection, we see that the real axis from $-\infty$ up to 0 is part of a locus, whereas there is no locus on the real axis from 0 to ∞. Therefore, the pole starting at 0 will go off to $-\infty$ on the real axis. (Remember also that the loci are always symmetric about the real axis, so a single pole can never break away from the real axis alone.)

8. There are no poles breaking away from the real axis, hence no breakaway point.

9. Looking at the direction of the asymptotes, it is clear that two loci will cross over into the positive real half plane. We can find the crossing point of the loci on the imaginary axis by finding the value of K which brings the system to this stability limit. The closed-loop characteristic equation is $1 + KG(s)$ (see the table in Section 4.1), and rearranging into standard form we have:

$$s^3 + 15s^2 + 90s + 100K = 0.$$

Using the Routh-Hurwitz criterion, it can be found that for stability, $K < 13.5$. Hence, for a value of $K = 13.5$, the poles will lie *on* the imaginary axis. We substitute $s = j\omega$ (remember that $\sigma = 0$ at this point) and $K = 13.5$ into the characteristic equation:

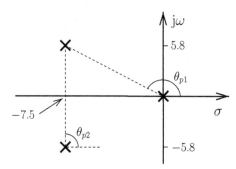

FIGURE 5.13 Sketch to determine departure angles.

$$(j\omega)^3 + 15(j\omega)^2 + 90j\omega + 1350 = 0$$
$$-j\omega^3 - 15\omega^2 + 90j\omega + 1350 = 0.$$

Equating real parts, we have:

$$-15\omega^2 + 1350 = 0 \Rightarrow \omega \approx \pm 9.5.$$

Hence, the loci will cross the imaginary axis at ± 9.5.

10. The loci will have departure angles from the complex open-loop poles, i.e., their initial direction as K increases from 0 upward. We make a simple sketch (see Figure 5.13) to find the required angles, as described in point 10 above (page 107).

We know that $\theta_{p2} = 90°$. Using standard trigonometry, we have that

$$\theta_{p1} = 180° - \arctan\left(\frac{5.8}{7.5}\right) \approx 142°.^5$$

Hence, the departure angle from this pole is

$$\theta_d = 180° - \theta_{p1} - \theta_{p2} = 180° - 142° - 90° = -52°.$$

The departure angle from the second complex pole can be calculated similarly; however, we know that the loci are symmetric about the real axis; therefore, by inspection we see that the departure angle from this pole must be 52°.

We can now finalize the sketch. The first branch is straightforward; it starts at 0 and follows the real axis to $-\infty$. The branches from the complex poles will

---

[5] Take care when finding tangents on the calculator so that you end up in the correct quadrant. Simply typing in $\arctan\left(\frac{5.8}{-7.5}\right)$ will often give an angle in the second quadrant, not the fourth.

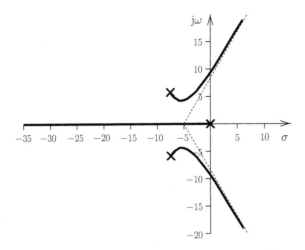

FIGURE 5.14 Example 2, final root locus sketch.

leave these at angles of $\mp52°$, pass through ±9.5 on the imaginary axis, and tend toward the asymptotes as K goes larger. Our final sketch is shown in Figure 5.14.

Now that we have drawn the root locus for our system, we can see the effect of increasing gain. In this case, increasing the gain will bring two closed-loop poles closer to the y-axis, meaning less damping and less stability. Initially, increasing the gain reduces the frequency (the value on the y-axis) slightly, but then the frequency increases again as the poles cross over into the positive half plane, i.e., the system goes unstable.

5.2.3 Root locus in Scilab and Matlab

Accurate root locus plots can be generated very easily with software packages such as Matlab and Scilab. Generating a root locus plot in Scilab is straightforward:

```
s=%s;
G = 100 / (s*(s^2+15*s+90)); // Open loop TF
evans(G); // Create root locus plot
```

For Matlab, this code will produce the same output:

```
s=tf('s');
G = 100 / (s*(s^2+15*s+90)); % Open loop TF
rlocus(G); % Create root locus plot
```

Note that the software automatically adds the gain variable, hence the open-loop transfer function must be defined without the variable K. The plot is shown in Figure 5.15a. This is actually the root locus for the system in the last example above, hence we can double-check that our sketch is reasonably close to the actual diagram. In the same way, we can double-check the sketch made in Example 1 above; the accurate root locus plot for this case is shown in Figure 5.15b.

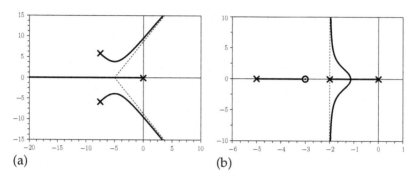

FIGURE 5.15 Example of root locus plots created with software packages. (a) Root locus plot for $G(s) = \frac{100}{s(s^2+15s+90)}$. (b) Root locus plot for $G(s) = \frac{(s+3)}{s(s+2)(s+5)}$.

A further point to note is that the software tools automatically adjust the range on the axes to what they think is appropriate, hence the plot may look different to that from a sketch. This can be adjusted in the figure properties window, and this is particularly important if we want to double-check any calculated departure or arrival angles. For the angle to show correctly, the steps on the x and y axes must be equal.

Finally, the `evans()` and `rlocus()` functions can take an argument to define the maximum value of K to be used. In theory, the root locus plot exists for K up to infinity, but we are most often only interested in what happens for reasonably low K-values. We can therefore add the maximum K to the command:

```
s=%s;
G = 100 / (s*(s^2+15*s+90)); // Open loop TF
evans(G,20); // Create root locus plot up to K=20
```

and

```
s=tf('s');
G = 100 / (s*(s^2+15*s+90)); % Open loop TF
rlocus(G,20); % Create root locus plot up to K=20
```

This gives a plot with better resolution for the sections we are most interested in.

5.2.4 Controller design using root locus

We have seen that the pole placement in the s-domain influences the dynamic behavior of the system. The information from the root locus plot can therefore help determine a suitable controller tuning. In this section, we will briefly recap the influence of the pole placement on the dynamic response of a system, and look at what information we can get from the root locus plot.

This section will look at second-order systems. Note that for a high-order plant, all the poles influence the system behavior; however, often there will be a set of dominant poles (the poles closest to the imaginary axis; see page 61) which will have the largest influence on the system response. By analyzing the locus of these poles, we can get a general idea of the system response for a given change in K. (Remember, it is only the principal system behavior we are after, not fine-tuning of the controller.)

SPEED OF RESPONSE AND OSCILLATING FREQUENCY

We have seen that the value of σ determines the speed of response of the system, since $e^{\sigma t}$ defines the converging (or diverging) envelope of the time-domain response. (See Figure 3.7 on page 50.) Hence, any poles moving toward the left-hand side in the pole-zero map will contribute to faster system response.

Figure 5.16 shows the time response of a second-order system for three pole positions. For the rightmost pole pair, $\sigma = -1$ and the response is slow and oscillatory. As the poles move toward the left, the response becomes quicker and the system settles faster; the lower value of σ leads to the oscillations dying out quicker. Note that because the value on the $j\omega$ axis is constant, the frequency of the response does not change.

Also, we know that the limit of stability is at $\sigma = 0$, hence any moving of the poles toward the right-hand side will bring the system closer to the unstable region. Similarly, poles moving toward the left-hand side can be said to make the system "more stable."

Similarly, we know from Chapter 3 that the value of ω determines the oscillating frequency of the system response. Figure 5.17 shows the effect of varying the imaginary component of the poles; the higher value of ω gives a higher oscillating frequency in the response. Also note that because σ is constant, the settling time is the same in all cases. Hence, when interpreting a root locus plot, we know that any changes in K leading to an increase in the imaginary parts of the poles lead to an increase in the frequency of the response, and vice versa.

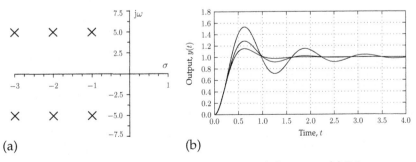

(a) (b)

FIGURE 5.16 Pole placement influence on system speed of response. (a) Pole-zero map. (b) Time response.

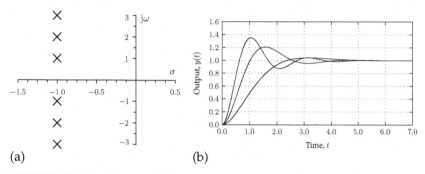

(a)   (b)

FIGURE 5.17 Pole placement influence on system speed of response. (a) Pole-zero map. (b) Time response.

DAMPING: GENERAL CASE FOR A SECOND-ORDER SYSTEM

A second-order system in standard form has a characteristic equation $s^2 + 2\zeta\omega_n s + \omega_n^2 = 0$, and if $\zeta < 0$, the system is underdamped and the poles are a complex conjugate pair. The roots for this system are:

$$s_1, s_2 = -\zeta\omega_n \pm j\,\omega_n\sqrt{1 - \zeta^2}.$$

A pole p_1 can then be represented in the pole-zero map as shown in Figure 5.18a.

Using standard trigonometry, it can be shown that $\cos\beta = \zeta$. Hence, the line from the origin through p_1 forms a line of constant damping ratio. This allows us to define a region of acceptable system damping by requiring that the poles lie within a specified region on the pole-zero map.

Within the shaded region in Figure 5.18b, the system will have a damping ratio equal to or less than $\cos\beta$. For example, in order to obtain a damping ratio of 0.5 or less, the system poles need to lie inside a region defined by $\beta = 60°$.

Figure 5.19a shows the effect of poles in a second-order system moving along the constant $\cos\beta$ line, and Figure 5.19b shows the corresponding time response.

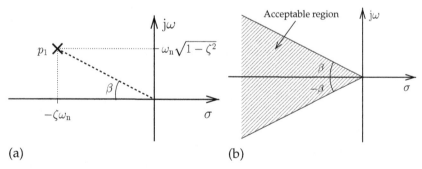

(a)   (b)

FIGURE 5.18 Pole placement influence on damping in Argand diagram. (a) System pole in Argand diagram. (b) Region of acceptable system damping ratios.

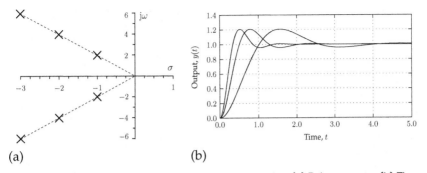

FIGURE 5.19 Pole placement influence on system damping. (a) Pole-zero map. (b) Time response.

It can be seen that the overshoot remains constant with constant damping ratio. Further, we are often interested in how large the overshoot of a system is in response to a change in the input. The relation between the percentage overshoot, m_{os}, and the damping ratio, ζ, is

$$\zeta = \frac{\left|\ln\left(\frac{m_{os}}{100}\right)\right|}{\sqrt{\pi^2 + \ln^2\left(\frac{m_{os}}{100}\right)}}$$

Using this, we can derive a table as shown in Table 5.1 for the relation between the angle of the acceptable region, damping ratio, and system overshoot. Hence, for interpreting a root locus plot, we know that if the angle β increases along a section of the root locus, increasing K will give less damping, i.e., more overshoot.

TABLE 5.1 Relation Between Pole Placement, Damping Ratio, and System Overshoot

β	ζ	m_{os} (%)
30°	0.87	1
45°	0.71	5
54°	0.59	10
60°	0.50	16
63°	0.46	20
66°	0.40	25
69°	0.36	30
73°	0.30	37

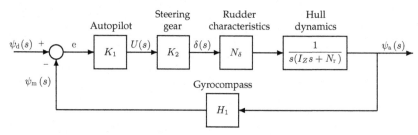

FIGURE 5.20 Block diagram of ship autopilot control system.

5.2.5 Example of ship autopilot system

Consider the ship autopilot example from above (Section 4.5.2). The block diagram is shown in Figure 5.20. The open-loop transfer function, which is the one we need to use for root locus analyses, for this system is

$$G(s) = K_1 \cdot K_2 \cdot N_\delta \cdot \frac{1}{s(I_Z s + N_r)} \cdot H_1$$

(Note that the open-loop transfer function includes any block in the feedback loop (in this case H_1).)

K_1 is the gain to be varied. Using the coefficient values from above, we get

$$G(s) = K_1 \cdot 1 \cdot 80 \times 10^6 \cdot \frac{1}{s(20 \times 10^9 s + 2 \times 10^9)} \cdot 1$$

$$= \frac{80 \times 10^6 \cdot K_1}{20 \times 10^9 s^2 + 10^9 s} = \frac{K_1}{250s^2 + 25s}$$

The root locus plot for this system is shown in Figure 5.21a. From the root locus we can read out some of the system characteristics. The system starts off

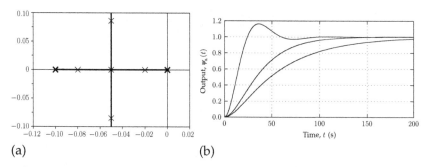

(a) (b)

FIGURE 5.21 Root locus plot, pole placement, and corresponding system time response. (a) Root locus plot. (b) Time response.

underdamped for low gain values, i.e., both poles are real. At some gain value, the poles meet at -0.05 and then moves into the imaginary plane, making the system response oscillatory. As the gain increases further, the poles move along the vertical asymptotes, hence the system will remain stable also for high gain values. (The poles do not cross into the positive, unstable, half plane.)

Figure 5.21b shows the time response to a step change in the input ψ_d for three gain values: $K_1 = 0, 4; 0.625; 2.5$. The corresponding pole placement is also superimposed on Figure 5.21a. In the three cases, we can see that:

- For $K_1 = 0.4$, the poles are at -0.02 and -0.08, and the system response is overdamped and very slow, only reaching a value near the steady-state response after around 200 s.
- For $K_1 = 0.625$, we have critical damping, i.e., the quickest possible response that can be achieved without overshoot. In this case there is a double pole at -0.05.
- A gain of $K_1 = 2.5$ gives complex poles at $-0.05 \pm j0.0866$ and a response which is oscillatory, however, with a significantly faster initial response than the previous two cases. Hence, if the overshoot can be accepted, this probably provides the best controller tuning of the three.

The plots shown in Figure 5.21 can be produced in Scilab with the following code (note that for the step response plots we have calculated the closed-loop transfer function, G_{cl}):

```
s=%s;
Gopen=syslin('c',1/(250*s^2+25*s));
t=0:0.01:200;
for K=[0.4,0.625,2.5]
    Gcl=syslin('c',K/(250*s^2+25*s+K));
    y=csim('step',t,Gcl);
    figure(1);
    plot(t,y);
    figure(2);
    plzr(Gcl);
end
evans(Gopen);
```

The equivalent code for Matlab is:

```
s=tf('s');
Gopen=1/(250*s^2+25*s);
for K=[0.4,0.625,2.5]
    Gcl=K/(250*s^2+25*s+K);
    figure(1);
    step(Gcl);
    figure(2);
    pzmap(Gcl);
end
rlocus(Gopen);
```

QUESTIONS

5.1 Use the Routh-Hurwitz stability criterion to show that the velocity feedback controller on a standard second-order plant (see Section 4.3.4) is always stable.

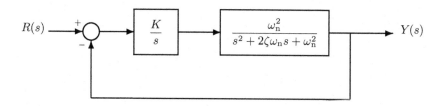

5.2 For the integral controller on a second-order plant in the last chapter, we saw that increasing the gain too much will bring the system to instability. (a) Assume that $\zeta = 1$ and $\omega_n = 1$ and use the Routh-Hurwitz stability criterion to find the values of K that will maintain stability for the closed-loop system. (b) Do the values of ζ and ω_n influence the maximum gain that can be used?

5.3 Referring back to the marine engine governor example in the questions of Chapter 4, the system is as follows:

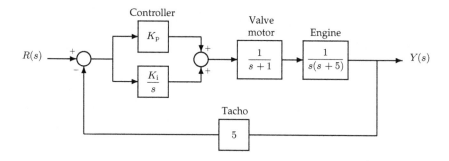

The forward path transfer function (i.e., not including the tacho) is

$$G_{fw}(s) = \left(K_p + \frac{K_i}{s}\right)\left(\frac{1}{s+1}\right)\left(\frac{1}{s(s+5)}\right) = \frac{sK_p + K_i}{s^2(s+1)(s+5)}.$$

The total feedback transfer function is

$$\frac{Y(s)}{R(s)} = \frac{\frac{sK_p + K_i}{s^2(s+1)(s+5)}}{1 + 5\left(\frac{sK_p + K_i}{s^2(s+1)(s+5)}\right)} = \frac{sK_p + K_i}{s^2(s+1)(s+5) + 5(sK_p + K_i)}.$$

Using the Routh-Hurwitz stability criterion, determine the range of values of K_p and K_i for which the system is stable.

5.4 For the system below, $K_p = K_i = 6$ and the element C is a gain factor.

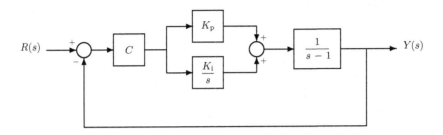

Find the closed-loop system transfer function. Determine the location of the closed-loop poles for $C = 1$ and $C = 0.5$ and mark them on the root locus below.

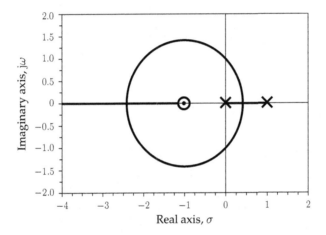

5.5 Draw the root locus for the system

$$G(s) = \frac{1}{s(s+4)(s+0.1)}.$$

Determine the value of gain for which the system is on the limit of stability. Now put a compensator in the forward path to introduce a pole at -8 and a zero at -1. Write down the transfer function of the compensator. Again, determine the gain at the limit of stability. What difference has the compensator made?

5.6 An unstable first-order system has a pole at $+1$ as shown in Figure 5.22.

It is modified by a PI controller with a transfer function $K_p + K_i/s$. With $K_p = 4$ and $K_i = 12$, the open-loop zero is at -3.

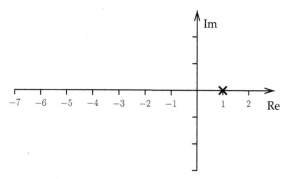

FIGURE 5.22 Pole-zero diagram for unstable first-order system.

(a) What value of K_p is required to put the zero at -4?

(b) Write out the total system and controller closed-loop transfer function for the new value of K_p.

Chapter Six

Frequency Domain Analysis

CHAPTER POINTS

- Frequency response
- Bode diagrams
- Gain and phase margin, bandwidth
- Nyquist diagrams
- *M* and *N* circles
- Nichols charts

6.1 FREQUENCY RESPONSE

Up until now we have looked at inputs mainly of the impulse, step, and ramp type (see Section 2.3.2). However, if we were to drive a system with a *sinusoidal* input, we would get a completely different response.[1] Consider for example, the tank heating system from Section 4.2.2, shown in Figure 6.1; if the valve signal *R* is a sine wave, the output from the tank will also be in the form of a sinusoid. Clearly, the output *Y* (measured temperature) from the tank will be a lot different if the input varies with a period of, for example, 1 h compared to 1 s, in terms of the tank's ability to respond to the demanded input.

When the input to a linear time-invariant system is a sinusoidal signal

$$u(t) = a \sin \omega t,$$

where *a* is a constant, the steady-state response is also a sine wave of the same frequency, ω, but having a different amplitude, *b*, and a phase shift, ϕ:

$$x(t) = b \sin(\omega t + \phi).$$

This is illustrated in Figure 6.2.

---

[1] For many engineering systems this would obviously not be done under normal operation, however, this assumption allows us to develop useful methods to analyze any dynamic system, as will be shown below.

Marine Systems Identification, Modeling, and Control. http://dx.doi.org/10.1016/B978-0-08-099996-8.00006-9

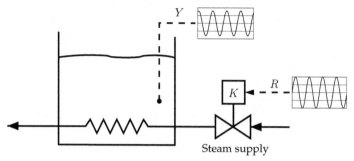

FIGURE 6.1 Tank heating system with sinusoidal input.

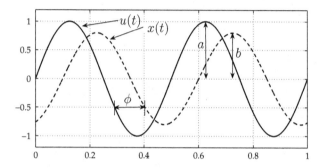

FIGURE 6.2 System response with a sinusoidal input signal.

6.1.1 Mathematical representation

In order to examine the frequency response characteristics of a system we recall the definition of the Laplace transform:

$$F(s) = \mathcal{L}\{f(t)\} = \int_0^\infty f(t)e^{-st}dt,$$

and the definition of the Fourier transform

$$F(j\omega) = \mathcal{F}\{f(t)\} = \int_{-\infty}^\infty f(t)e^{-j\omega t}dt.$$

When $f(t)$ exists for $t \geq 0$ we can see that the two transforms are closely related, i.e., to "move" from Laplace to Fourier we can substitute $s = j\omega$. Thus we can write:

$$G(s) = G(j\omega) = R(j\omega) + jI(j\omega)$$

where

$$R(j\omega) = \text{Re}(G(j\omega)) \quad \text{and}$$
$$I(j\omega) = \text{Im}(G(j\omega)).$$

Alternatively,

$$G(j\omega) = |G(j\omega)| e^{j\phi(j\omega)}$$
$$= |G(j\omega)| \angle \phi(j\omega)$$

where

$$\phi(j\omega) = \tan^{-1}\{I(j\omega)/R(j\omega)\} \quad \text{and}$$
$$|G(j\omega)|^2 = (R(j\omega))^2 + (I(j\omega))^2.$$

In other words, to determine the frequency response of a system described by its transfer function, we substitute s for $j\omega$ and calculate the modulus and argument. (See also Appendix B.1.)

Thus, the frequency response for the standard first-order function $G(s) = \dfrac{K}{1 + \tau s}$ can be found as:

$$|G(j\omega)| = \left| \frac{K}{1 + \tau j\omega} \right| = \frac{K}{\sqrt{1 + \tau^2 \omega^2}}$$
$$\phi(j\omega) = \tan^{-1}\left(\frac{0}{K}\right) - \tan^{-1}\left(\frac{\tau\omega}{1}\right)$$
$$= -\tan^{-1}(\tau\omega).$$

For a second-order system in standard form, $G(s) = \dfrac{K\omega_n^2}{s^2 + 2\zeta\omega_n s + \omega_n^2}$, we have a modulus:

$$|G(j\omega)| = \left| \frac{K\omega_n^2}{j^2\omega^2 + 2\zeta\omega_n j\omega + \omega_n^2} \right|$$
$$= \frac{K\omega_n^2}{\sqrt{\omega_n^4 + \omega^4 + \omega_n^2\omega^2(4\zeta^2 - 2)}},$$

and an argument

$$\phi(j\omega) = \tan^{-1}\left(\frac{0}{K\omega_n^2}\right) - \tan^{-1}\left(\frac{2\zeta\omega_n\omega}{\omega_n^2 - \omega^2}\right)$$
$$= -\tan^{-1}\left(\frac{2\zeta\omega_n\omega}{\omega_n^2 - \omega^2}\right).$$

Normally, $|G(j\omega)|$ is written as M, the magnitude, together with ϕ, the phase.

The frequency response can be depicted in a variety of ways, the most common of which are Cartesian plots as shown in Figure 6.3a. In a slightly modified form, this method of displaying frequency responses will be seen later.

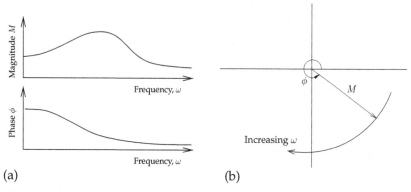

FIGURE 6.3 Plots of system frequency response. (a) Cartesian plot of system frequency response. (b) Polar plot of system frequency response.

The other common method is to use polar plots as in Figure 6.3b. (Note that the same result can be obtained by plotting $R(j\omega)$ against $I(j\omega)$.)

6.2 STABILITY IN THE FREQUENCY DOMAIN

Consider the standard system shown in Figure 6.4, but which has an additional negative gain term in the feedback loop. In the original configuration, i.e., without the -1 term, this system is always stable. However, the introduction of the negative gain changes the closed-loop transfer function to

$$\frac{Y(s)}{R(s)} = \frac{\dfrac{k_c k_p}{1 + \tau s}}{1 + \dfrac{-k_c k_p k_f}{1 + \tau s}} = \frac{k_c k_p}{\tau s + 1 - k_c k_p k_f}.$$

This system has a pole at $\dfrac{k_c k_p k_f - 1}{\tau}$. We know that τ is always positive, so for the pole to be in the left half plane, i.e., for the system to be stable, it is clear that $k_c k_p k_f < 1$.

Now, if an open-loop system has a phase shift of 180°, that would be equivalent to a negative unit gain as in the case above, since $\sin(\alpha + 180) = -\sin(\alpha)$.

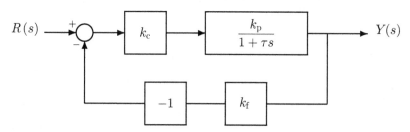

FIGURE 6.4 Positive feedback system.

Hence, if an open-loop system has a phase shift of 180°, the open-loop gain ($k_c k_p k_f$) must be *less than* that for the *closed-loop* system to be stable.

Frequency response methods therefore provide a useful tool for analyzing system stability. Importantly, these methods not only determine simply whether a system is stable or not (like the Routh Hurwitz criterion does) but also *how stable* it is. This is done by determining how much the gain must be increased, or how much the phase must shift, to reach instability. This information can be found directly in frequency response plots, as we will see later.

6.3 BODE DIAGRAMS

We have seen the use of Cartesian plots for frequency response; however, using logarithmic plots results in a powerful design tool. The advantage of logarithmic plots is that the mathematical operations of multiplication and division are transformed to addition and subtraction. Furthermore, any complex transfer function can be constructed from four basic types of factors and these can be easily plotted by means of straight-line approximations. These allow the performance characteristics of a control system to be obtained very quickly.

In constructing Bode diagrams rather than plot $\log_{10}|G(j\omega)|$, i.e., the logarithm of the magnitude directly, it is converted into decibels. This is called the log magnitude, abbreviated L_M. Hence,

$$L_M\{G(j\omega)\} = 20\log_{10}|G(j\omega)| \quad [\text{dB}].$$

Thus a Bode diagram comprises two parts, the log magnitude and the phase plotted against input frequency, ω, as shown in Figure 6.5. Further, the frequency is plotted on a logarithmic scale; this allows us to sketch Bode plots using straight-line approximations, as will be shown below.

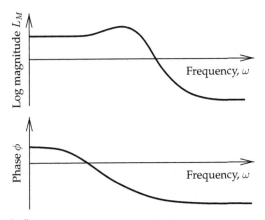

FIGURE 6.5 Bode diagram.

6.3.1 Generalized transfer function

Any transfer function can be written in a generalized form using a set of known factors:

$$G(s) = \frac{K_M(1 + \tau_1 s)(1 + \tau_2 s)^m \ldots}{(s)^n(1 + \tau_a s)\left(1 + \left(\frac{2\zeta}{\omega_n}\right)s + \left(\frac{1}{\omega_n^2}\right)s^2\right)\ldots},$$

which can be transformed into the frequency domain by letting $s = j\omega$:

$$G(j\omega) = \frac{K_M(1 + \tau_1 j\omega)(1 + \tau_2 j\omega)^m \ldots}{(j\omega)^n(1 + \tau_a j\omega)\left(1 + \left(\frac{2\zeta}{\omega_n}\right)j\omega + \left(\frac{1}{\omega_n^2}\right)(j\omega)^2\right)\ldots}.$$

Thus the log magnitude is given by

$$\begin{aligned}L_M\{G(j\omega)\} = {} & L_M\{K_M\} + L_M\{(1 + j\omega\tau_1)\} + mL_M\{(1 + j\omega\tau_2)\} - \\ & nL_M\{(j\omega)\} - L_M\{(1 + \tau_a j\omega)\} - \\ & L_M\left\{\left(1 + \left(\frac{2\zeta}{\omega_n}\right)j\omega + \left(\frac{1}{\omega_n^2}\right)(j\omega)^2\right)\right\}\ldots\end{aligned}$$

and the phase angle is given by

$$\begin{aligned}\phi = {} & \tan^{-1}\left(\frac{0}{K_M}\right) + \tan^{-1}\left(\frac{\omega\tau_1}{1}\right) + m\tan^{-1}\left(\frac{\omega\tau_2}{1}\right) \\ & - n\tan^{-1}\left(\frac{\omega}{0}\right) - \tan^{-1}\left(\frac{\omega\tau_a}{1}\right) - \tan^{-1}\left(\frac{2\zeta\frac{\omega}{\omega_n}}{\left(1 - \frac{\omega^2}{\omega_n^2}\right)}\right)\ldots\end{aligned}$$

It is clear that there are four basic types of factors:

1. K_M
2. $(j\omega)^{\pm n}$
3. $(1 + j\omega\tau)^{\pm m}$
4. $\left(1 + \left(\frac{2\zeta}{\omega_n}\right)j\omega + \left(\frac{1}{\omega_n^2}\right)(j\omega)^2\right)^{\pm p}$

Thus if we can draw the Bode diagrams for these individually, then by "graphical" addition it is possible to construct the Bode diagram for any complex system.

1. Constant factors

The constant K_M is independent of frequency, thus

$$L_M\{K_M\} = 20\log_{10}K_M$$

is a horizontal straight line and the phase angle is zero ($\phi = \tan^{-1}(0/K_M)$). This gives a Bode diagram for constant factors as shown in Figure 6.6.

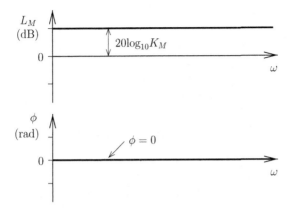

FIGURE 6.6 Bode diagram for constant factors.

2. $j\omega$ factors

For $j\omega$ factors appearing in the denominator, the log magnitude is given by

$$L_M \left\{ \frac{1}{j\omega} \right\} = 20 \log_{10} \left| \frac{1}{j\omega} \right| = -20 \log_{10} \omega.$$

When plotted against ω, the result is a straight line with a negative slope of 20 dB/decade. The phase angle ϕ is given by $\phi = -\tan^{-1}\left(\frac{\omega}{0}\right) = -\frac{\pi}{2}$, i.e., the phase is constant at $-90°$. The resulting Bode plot is shown in Figure 6.7a.

When the $j\omega$ terms appear in the numerator, the log magnitude is

$$L_M \{j\omega\} = 20 \log_{10} |j\omega| = 20 \log_{10} \omega$$

which again is a straight line but with a positive slope of 20 dB/decade. The phase angle is

$$\phi = \tan^{-1}\left(\frac{\omega}{0}\right) = \frac{\pi}{2}.$$

The Bode plot is shown in Figure 6.7b.

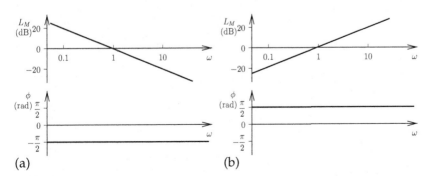

FIGURE 6.7 Bode diagram for $j\omega$ factors. (a) $j\omega$ factors in the denominator. (b) $j\omega$ factors in the numerator.

3. $(1 + j\omega\tau)$ factors

A factor $(1 + j\omega\tau)$ appearing in the denominator has a log magnitude

$$L_M \left\{ \frac{1}{1 + j\omega\tau} \right\} = 20 \log_{10} \left| \frac{1}{1 + j\omega z} \right| = -20 \log \sqrt{1 + \omega^2 \tau^2}$$

For small values of ω, that is $\omega\tau \ll 1$, this expression approximates to $-20 \log_{10} 1 = 0 \, \text{dB}$. Thus for low frequencies, the plot of log magnitude is the $0 \, \text{dB}$ line.

For large values of ω, that is $\omega\tau \gg 1$, the log magnitude is

$$L_M \left\{ \frac{1}{1 + j\omega\tau} \right\} = 20 \log_{10} \left(\frac{1}{j\omega\tau} \right) = -20 \log_{10} \omega\tau.$$

When $\omega = \dfrac{1}{\tau}$, the log magnitude is $L_M \left\{ \dfrac{1}{1 + j\omega\tau} \right\} = 0 \, \text{dB}$. From the above, we see that when $\omega > \dfrac{1}{\tau}$, the magnitude can be approximated as a straight line with a negative slope of $20 \, \text{dB/decade}$. Thus the log magnitude plot for $(1 + j\omega\tau)$ factors in the denominator will look like the one shown in Figure 6.8a.

The frequency at which $\omega = 1/\tau$ is called the corner frequency. The physical interpretation of this diagram is that up to the corner frequency, the steady-state output will have the same magnitude as the command input. Thereafter, the steady-state output magnitude decreases for a constant command input magnitude.

The phase angle is given by

$$\phi = \tan^{-1} \left(\frac{1}{1 + j\omega\tau} \right) = -\tan^{-1}(\omega\tau).$$

For small values of ω, the phase angle will be zero, while for large values of ω, the phase angle will be $-\pi/2$. A straight-line approximation for the phase angle can be drawn as shown in Figure 6.8b. We assume that up to one decade below the corner frequency, the phase angle is zero, at one decade higher than the

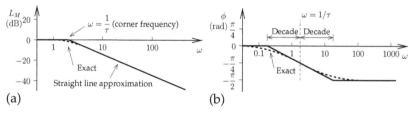

FIGURE 6.8 Bode diagram for $(1 + j\omega\tau)$ factors in the denominator. (a) Log magnitude plot. (b) Phase plot.

corner frequency, the phase angle has reached $-\pi/2$, and that the phase angle drops linearly between these two points.

When the factor $(1 + j\omega\tau)$ appears in the numerator, it has the log magnitude and phase

$$L_M = 20 \log \sqrt{1 + \omega^2\tau^2}$$
$$\phi = \tan^{-1}(\omega\tau).$$

In other words, we will have the mirror image (about the x-axis) of the denominator version shown above. Again, straight-line approximations can be drawn.

4. Quadratic factors

When the damping ratio is greater than or equal to 1, $\zeta \geq 1$, the quadratic term can be factored into two first-order factors which can be plotted separately as shown above. However, for $\zeta < 1$, the quadratic factor becomes complex. For denominator factors, we get:

$$L_M = -20 \log_{10} \left(\left(1 - \frac{\omega^2}{\omega_n^2}\right)^2 + \left(\frac{2\zeta\omega}{\omega_n}\right)^2 \right)^{\frac{1}{2}}$$

and

$$\phi = -\tan^{-1} \left(\frac{2\zeta\omega/\omega_n}{(1 - \omega^2/\omega_n^2)} \right).$$

For low frequencies, $L_M \approx 0\,\text{dB}$, while at high frequencies we have

$$L_M \approx -20 \log_{10} \frac{\omega^2}{\omega_n^2} \approx -40 \log_{10} \left(\frac{\omega}{\omega_n}\right).$$

This means at high frequencies, the frequency response plot will approach an asymptote which has a negative slope of 40 dB/decade.

From a straight-line approximation, as shown in Figure 6.9a, we can see that at low frequencies the output of the system follows the command input

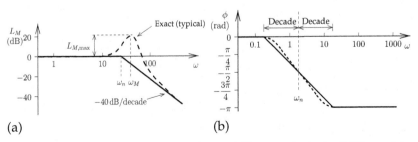

(a)   (b)

FIGURE 6.9 Bode diagram for quadratic factors. (a) Log magnitude plot. (b) Phase plot.

and from around the undamped natural frequency, ω_n, the output declines by 40 dB/decade. However, as the input frequency nears the system resonant frequency, ω_n, the output has a greater magnitude, reaching a maximum at $\omega = \omega_M$. Thereafter, the output magnitude decreases toward the asymptotic value of 40 dB/decade.

If we consider the magnitude for a quadratic expression, we can write

$$M^2 = \frac{1}{\left(1 + \frac{\omega^2}{\omega_n^2}\right)^2 + 4\zeta^2 \frac{\omega^2}{\omega_n^2}}.$$

The maximum value will occur when

$$\frac{dM^2}{d\omega} = \frac{-4\left(1 - \frac{\omega^2}{\omega_n^2}\right)\left(\frac{\omega}{\omega_n}\right)^2 + 8\zeta^2 \left(\frac{\omega^2}{\omega_n^2}\right)^2}{\left(\left(1 - \frac{\omega^2}{\omega_n^2}\right)^2 + 4\zeta^2 \left(\frac{\omega^2}{\omega_n^2}\right)\right)^2} = 0$$

The maximum value is

$$M_{\max} = \frac{1}{2\zeta\sqrt{1 - \zeta^2}},$$

and this occurs at the frequency

$$\omega_M = \omega_n\sqrt{1 - 2\zeta^2}.$$

Clearly, a "resonant peak" will only appear if $M > 1$, that is

$$\frac{1}{2\zeta\sqrt{1 - \zeta^2}} > 1,$$

which occurs if $\zeta < 0.707$.

To plot it in the Bode diagram, the log magnitude of the maximum is needed, and is given by

$$L_{M_{\max}} = 20\log_{10}\left(\frac{1}{2\zeta\sqrt{1 - \zeta^2}}\right) \quad \text{for } \zeta < 0.707$$

The phase diagram is zero for low ω and $-\pi$ for high ω, and a straight-line approximation can be created as above. This is shown in Figure 6.9b.

6.3.2 Drawing complex Bode diagrams

We now have all the basic diagrams to construct complex Bode diagrams. Consider the transfer function

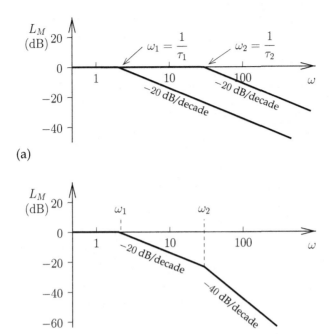

FIGURE 6.10 Drawing complex Bode plots. (a) Straight-line approximations for the two first-order terms. (b) "Adding" these together gives the overall system Bode plot.

$$G(s) = \frac{1}{(1 + \tau_1 s)(1 + \tau_2 s)},$$

where $\tau_1 > \tau_2$.

The overall log magnitude is given by the sum of the contributions from each $(1 + j\omega\tau)$ term, i.e.,

$$L_M = L_{M1} + L_{M2}$$

L_{M1} and L_{M2} can be plotted easily using the previously shown straight-line approximations, as shown in Figure 6.10a. "Adding" these together gives the final Bode plot in Figure 6.10b.

A similar approach can be used for the phase plot.

6.3.3 Stability margins and bandwidth in the Bode plot

We are often interested in knowing "how stable" a system is, and this can be specified by the *gain margin* and the *phase margin*. Whether a system will be stable in the closed loop, and also the approximate degree of stability, can be determined directly from the Bode plot. The stability characteristics are specified in terms of:

- Gain crossover: This is the point on the plot at which the magnitude is unity: $L_M(G(j\omega)) = 0\,$dB. The frequency at the gain crossover is called the phase margin frequency ω_ϕ.
- Phase margin: This is 180° plus the negative trigonometrically considered phase of the transfer function at the gain crossover point. The phase margin can be considered as the amount of phase shift at the frequency ω_ϕ that would produce instability.
- Phase crossover: This is the point on the plot at which the phase is −180°. The frequency at which this occurs is called the gain margin frequency ω_c.
- Gain margin: Measured at the phase crossover frequency, the gain margin is the factor by which the gain must be changed in order to produce instability; a gain of more than 1 (0 dB) at this point would produce instability (see Section 6.2).

Gain and phase margins are useful measures of stability, as they will identify not only whether a system is stable or not for a given situation (which can be also found using Routh Hurwitz) but also indicate *how close* it is to instability. It is common to specify a gain margin of 6 dB and a phase margin of 45°. The reason for this is that the control system design may have been performed using a simplified mathematical model, and there may be factors not properly accounted for. These gain and phase margins ensure, in general, that any modeling errors will not cause the system to be unstable in practice.

Examples

For a system which is stable in the closed loop, shown in Figure 6.11a, the system gain at −180° of phase shift (i.e., at the phase crossover, ω_c) is less than 1 (0 dB). The gain margin is the amount the system gain can be increased before the system reaches the point of marginal stability.

For a system which is unstable in the closed loop, as shown in Figure 6.11b, there will exist an input frequency which will give a phase shift of −180° and an overall gain of more than 0 dB.

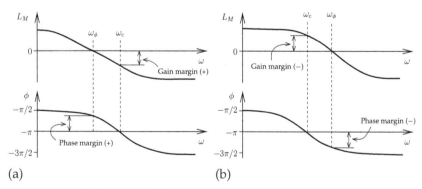

FIGURE 6.11 System stability in the Bode diagram. (a) Bode diagram for a stable system. (b) Bode diagram for an unstable system.

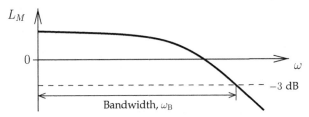

FIGURE 6.12 System bandwidth in the Bode diagram.

Bandwidth

One final and useful measure is the bandwidth, which can also be determined from a Bode diagram. It is important in many problems to consider the range of frequencies for which the system "gain" is adequate to ensure a good tracking performance.

The length of this range of frequencies is called the bandwidth, ω_B, and the gain is typically considered adequate if it is greater than 0.707 or -3 dB. This is illustrated in Figure 6.12.

6.3.4 Bode plots in Matlab and Scilab

Exact Bode plots can be generated very quickly using software tools. For example, the Bode plot for $G(s) = \dfrac{100}{s(s^2 + 15s + 90)}$ can be generated as shown below for Scilab (left) and Matlab (right).

```
s=%s;
G=syslin('c',100/(s*(s^2+15*s+90)));
bode(G,"rad");
```

```
s=tf('s');
G=100/(s*(s^2+15*s+90));
bode(G);
```

6.4 NYQUIST DIAGRAMS

A Nyquist diagram is a version of the polar plot format for frequency response. It is useful in that it provides a simple graphical procedure for determining the closed-loop stability from the frequency response curves of the open-loop transfer function $KG(s)$.

The closed-loop stability of standard systems can be evaluated in a simplified form (for a full derivation of the Nyquist stability criterion, see, e.g., Burns [1]), known as the "left-hand criterion": If $KG(s)$ has *no* poles or zeros having positive real parts, then $1 + KG(s) = 0$ has *no* unstable roots. Therefore, if the $KG(j\omega)$ plot is traced out as ω goes from 0^+ to $+\infty$ it will always leave the $(-1, 0)$ point on its left.

Example

The use of Nyquist diagrams with the left-hand criterion is most easily demonstrated by an example. Plot the Nyquist diagram of the following plant to determine its closed-loop stability:

$$G(s) = \frac{1}{(1 + 0.2s)(1 + s)(1 + 10s)},$$

for (a) $K = 10$, (b) $K = 136.8$, and (c) $K = 500$.

First we substitute for $j\omega$ and work out the open-loop transfer function:

$$KG(s) = \frac{K}{(1 + 0.2s)(1 + s)(1 + 10s)}$$

$$KG(j\omega) = \frac{K}{(1 + 0.2j\omega)(1 + j\omega)(1 + 10j\omega)}$$

$$KG(j\omega) = \frac{K}{(1 - 12.2\omega^2) + (11.2\omega - 2\omega^3)j}$$

This gives the following expressions for magnitude and argument:

$$M = \frac{K}{\sqrt{(1 - 12.2\omega^2)^2 + (11.2\omega - 2\omega^3)^2}}$$

$$\phi = \tan^{-1}\left(\frac{0}{K}\right) - \tan^{-1}\left(\frac{11.2\omega - 2\omega^3}{1 - 12.2\omega^2}\right)$$

Substituting and working out the magnitude and phase for a set of selected frequencies, we get values as in Table 6.1.

The resulting Nyquist plot is sketched in Figure 6.13. The three gain values chosen were not arbitrary but are in fact one value that gives a stable system in the closed loop ($K = 10$), one that gives a marginally stable system ($K = 136.8$), and one which would bring the system to instability if closing the loop ($K = 500$). (You can prove this with the techniques shown in the previous chapters, e.g., the Routh Hurwitz stability criterion.)

It can be seen that for a stable system, the Nyquist plot passes to the left of the -1 point on the real axis. At marginal stability, the plot passes *through* the -1 point, whereas for an unstable system, the Nyquist contour will encircle the -1 point.

TABLE 6.1 Magnitude and Phase Values for Nyquist Plot

ω	$K = 10$ M	ϕ	$K = 136.8$ M	ϕ	$K = 500$ M	ϕ
0	10	0	136.8	0	500	0
1	0.7	$-140.6°$	19.8	$-140.6°$	72.5	$-140.6°$
2	0.2	$-172.4°$	2.8	$-172.4°$	10.4	$-172.4°$
3	0.1	$-190.6°$	1.2	$-190.6°$	4.5	$-190.6°$
∞	0	$-270°$	0	$-270°$	0	$-270°$

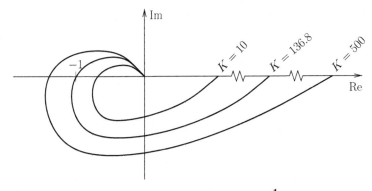

FIGURE 6.13 Sketch of Nyquist plot for $G(s) = \dfrac{1}{(1+0.2s)(1+s)(1+10s)}$.

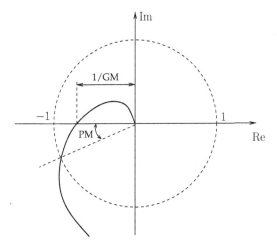

FIGURE 6.14 Determining gain and phase margin from Nyquist plots.

6.4.1 Stability margins in the Nyquist diagram

Similarly, as in the Bode plot, stability margins can be read off the Nyquist diagram directly, as shown in Figure 6.14. The point at which the Nyquist contour crosses the real axis is equal to the inverse of the gain margin. (Note that this may need to be converted into decibels if comparing it to a value read off a Bode plot.) The phase margin is the angle of the unity magnitude response to the real axis, i.e., the angle at the point where the magnitude crosses a unit circle.

6.5 *M* AND *N* CIRCLES

Having utilized the open-loop frequency response plots to determine system stability by using Nyquist's criterion, it would also be desirable if we could use the same plots to find the resulting frequency response for the closed-loop system. This can be done using M and N circles, this being the common terminology for the closed-loop amplitude and the closed-loop phase angle, respectively.

6.5.1 M circles

Consider the closed-loop transfer function

$$G_{CL}(s) = \frac{KG(s)}{1 + KG(s)}.$$

The closed-loop magnitude is given by

$$M(j\omega) = \frac{|KG(j\omega)|}{|1 + KG(j\omega)|}.$$

Consider now writing the open-loop transfer function in terms of rectangular coordinates: $KG(j\omega) = x + jy$. Substituting this gives

$$M = \frac{|x + jy|}{|1 + x + jy|} = \left(\frac{x^2 + y^2}{(x + 1)^2 + y^2} \right)^{1/2},$$

or

$$M^2 = \frac{x^2 + y^2}{(x + 1)^2 + y^2}.$$

This can be rearranged to give

$$\left(x + \frac{M^2}{M^2 - 1} \right)^2 + y^2 = \left(\frac{M}{M^2 - 1} \right)^2,$$

which is the equation of a family of circles depending upon the value of M. The center of the circle is given by

$$x_0 = \frac{-M^2}{M^2 - 1}, \quad y_0 = 0$$

and the radius is

$$r_0 = \left| \frac{M}{M^2 - 1} \right|.$$

This is illustrated in Figure 6.15. The M circles are to the left of $x = -0.5$ for $M > 1$, and to the right of $x = -0.5$ for $M < 1$. When $M = 1$, the circle becomes the straight line at $x = -0.5$.

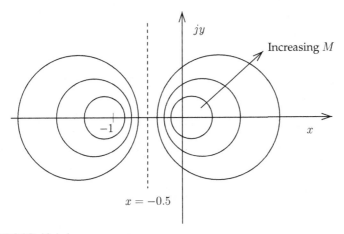

FIGURE 6.15 *M* circles.

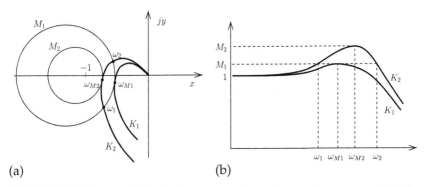

(a)   (b)

FIGURE 6.16 The use of *M* circles for evaluating closed-loop system response. (a) Polar plot. (b) Cartesian plot.

The use of *M*-circles can be seen from a simple example. Figure 6.16a shows the polar frequency response plot with superimposed *M* circles, and Figure 6.16b shows the corresponding Cartesian plot. As can be seen, it is possible to read off the values of $M_{\max}$ and ω_M directly from the polar frequency response.

6.5.2 *N* circles

In a similar manner, we can obtain circles of constant closed-loop phase angles. As above, we consider the closed-loop transfer function

$$G_{CL}(s) = \frac{KG(s)}{1 + KG(s)}.$$

Writing the frequency response in terms of rectangular coordinates, we get

$$KG(j\omega) = x + jy.$$

Then

$$G_{CL}(j\omega) = \frac{x + jy}{1 + x + jy} = Me^{j\theta}$$

and

$$\theta = \tan^{-1}\left(\frac{y}{x}\right) - \tan^{-1}\left(\frac{y}{1+x}\right)$$

$$\theta = \tan^{-1}\left(\frac{y}{x^2 + x + y^2}\right),$$

or

$$\tan\theta = \frac{y}{x^2 + x + y^2} = N.$$

For a constant angle θ, this equation can be rearranged to give

$$\left(x + \frac{1}{2}\right)^2 + \left(y - \frac{1}{2N}\right)^2 = \frac{N^2 + 1}{4N^2}.$$

This is also the equation of a circle with N as a parameter. The center is at

$$x = -\frac{1}{2}, \quad y = \frac{1}{2N},$$

and the circle will have a radius of

$$r = \frac{1}{2}\left(\frac{N^2 + 1}{N^2}\right)^{1/2}.$$

This is shown in Figure 6.17. Thus, it is possible to read closed-loop phase angles directly from the constant N-circles plot.

6.5.3 Nichols charts

Often it is more convenient to have M and N loci plotted in the gain-phase plane rather than the polar plane. Such a plot is called a Nichols chart, and is shown in Figure 6.18.

When the open-loop frequency response information is plotted on a Nichols chart, the closed-loop harmonic response information can be read off directly from intersections of the locus with the M and N contours.

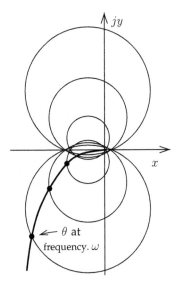

FIGURE 6.17 *N* circles.

For example, construct the magnitude-phase plot for a closed-loop frequency response of the unity feedback system whose open-loop transfer function is

$$G(s) = \frac{2}{s(1 + s)(1 + s/3)}.$$

First, we substitute $j\omega$ for s and hence find the magnitude and phase equations for the open-loop transfer function. This gives:

$$G(s) = \frac{2}{s(1 + s)(1 + s/3)}$$

$$G(j\omega) = \frac{6}{(-4\omega^2) + j(3\omega - \omega^3)}$$

$$M = |G(j\omega)| = \frac{6}{\sqrt{(-4\omega^2)^2 + (3\omega - \omega^3)^2}}$$

$$\phi = \angle G(j\omega) = -\tan^{-1}\left(\frac{3\omega - \omega^3}{-4\omega^2}\right)$$

The calculated open-loop L_M and ϕ values, shown in Table 6.2, can now be plotted on the Nichols chart. This is shown in Figure 6.19. Then, the corresponding L_M and ϕ values for the closed-loop system can be read off the chart by interpolation of the M and N contours of the Nichols chart. The readings are shown in Table 6.3.

Because this is a simple example it is relatively straightforward to calculate the closed-loop harmonic information directly. This will provide a check that the use of Nichols charts is valid.

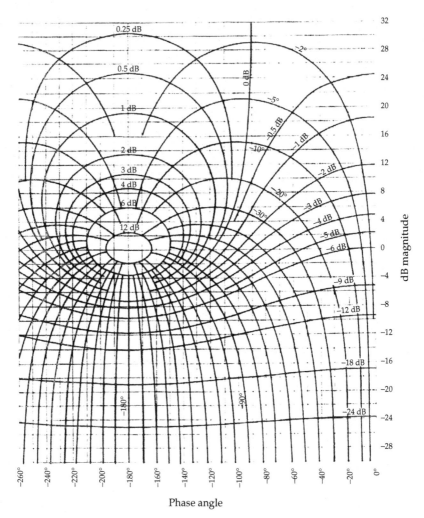

FIGURE 6.18 Nichols chart.

TABLE 6.2 Calculated Open-Loop L_M and ϕ Values

Frequency, ω	Magnitude M	L_M (dB)	Phase, ϕ (°)
0	∞	∞	
0.2	9.784	19.810	−105.12
0.5	3.529	10.953	−126.03
1.0	1.342	2.555	−153.43
1.25	0.923	−0.696	−163.96
1.5	0.662	−3.583	−172.87
2.0	0.372	−8.589	−187.13
3.0	0.149	−16.536	−206.57

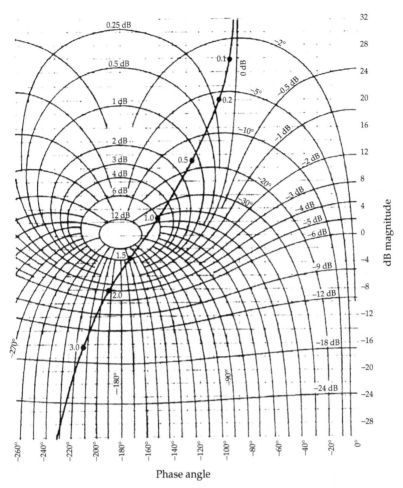

FIGURE 6.19 Nichols chart with open-loop L_M and ϕ values.

TABLE 6.3 Closed-Loop L_M and ϕ Values

Frequency, ω	Graphical		Calculated	
	Magnitude, L_M (dB)	Phase, ϕ (°)	Magnitude, L_M (dB)	Phase, ϕ (°)
0	0	0	0	0
0.2	0.2	−6	0.20	−5.8
0.5	1.2	−15	1.27	−15.4
1.0	6	−42	6.54	−45
1.25	10	−90	10.38	−97.9
1.5	6.0	−155	5.46	−159.5
2.0	−4.0	−194	−4.62	−191.3
3.0	−15.0	−212	−15.4	−210.5

The closed-loop transfer function and the corresponding expressions for magnitude and phase are:

$$\frac{G(j\omega)}{1 + G(j\omega)} = \frac{6}{(j\omega)^3 + 4(j\omega)^2 + 3j\omega + 6}$$

$$= \frac{6}{(6 - 4\omega^2) + j(3\omega - \omega^3)}$$

$$M = \frac{6}{\sqrt{(6 - 4\omega^2)^2 + (3\omega - \omega^3)^2}}$$

$$\phi = -\tan^{-1}\left(\frac{3\omega - \omega^3}{6 - 4\omega^2}\right)$$

The calculated results are also shown in Table 6.3. We see that these concur well with those obtained graphically, hence showing the validity of using the Nichols chart to obtain closed-loop harmonic information.

6.6 EXAMPLE: AUTONOMOUS UNDERWATER VEHICLE

Figure 6.20 shows an autonomous underwater vehicle (AUV) with control fins for heave/dive control. In the control system, the fin pitch angle δ is the input and the vessel heading angle θ is the output.

A proportional controller with gain K is used. The open-loop block diagram is shown in Figure 6.21.

The open-loop transfer function for this situation is

$$\frac{\theta(s)}{\delta(s)} = K\frac{2}{(s + 2)}\frac{0.125(s + 0.435)}{(s + 1.23)(s^2 + 0.226s + 0.0169)}.$$

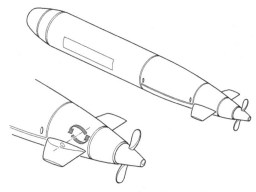

FIGURE 6.20 Autonomous underwater vehicle with control fins.

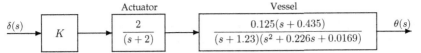

FIGURE 6.21 AUV open-loop control diagram.

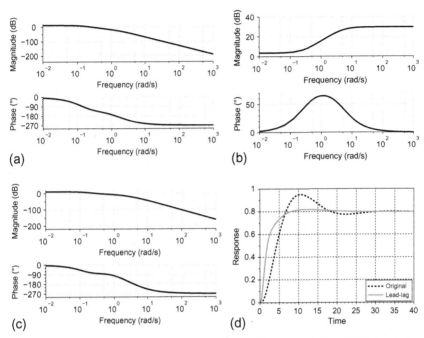

FIGURE 6.22 AUV system characteristics with two different controllers. (a) Bode plot for original system. (b) Bode plot for lead-lag controller. (c) Bode plot for system with the new controller. (d) Step response for the system with the two controllers.

In the original configuration, a controller gain of $K = 1.5$ is used. This gives a system gain margin of 24 dB and a phase margin of 62°. Figure 6.22a shows the Bode plot for this case.

In order to further improve the stability margins, it has been proposed to use a lead-lag compensator with a transfer function:

$$G_c(s) = \frac{29.655(s + 0.2585)}{(s + 5.11)}.$$

The Bode diagram for the lead-lag term is shown in Figure 6.22b. It can be seen that the contribution from the new controller will be a positive shift in the phase in the range around the gain crossover frequency, plus a gain increase for higher frequencies.

Figure 6.22c shows the system Bode plot with the new controller. It can be seen that the phase margin has been increased significantly by the additional lead-lag term, and is now 99°. The gain plot has been influenced to some degree; however, the phase crossover frequency ω_c has been shifted so that the gain margin with the new controller remains at 24 dB.

Figure 6.22d shows the response of the closed-loop system to a step input in the heading demand signal. It is seen that, in addition to improved stability, the lead-lag controller provides a significantly better time response, with a faster initial response and no oscillations.

This example shows how frequency response methods can be used when choosing or tuning a controller for a particular system; when knowing the frequency response of a plant, a suitable controller can be selected in order to obtain desirable system characteristics.

QUESTIONS

6.1 Match the transfer function to the bode plots below using appropriate straight-line approximation methods.

TF1	$\dfrac{1}{(s+1)}$	TF2	$\dfrac{s}{(s+1)(s+100)}$
TF3	$\dfrac{s}{(s+1)}$	TF4	$\dfrac{1}{(s+1)(s+100)}$

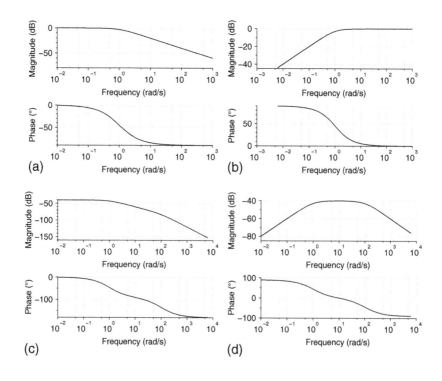

6.2 The hydraulic actuation system shown in the figure uses an error actuated proportional controller with a fixed gain of K_p.

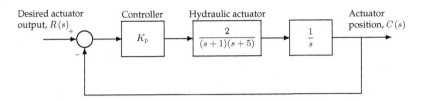

(a) With the proportional gain set at $K_p = 5$, draw the Bode diagram for this system and find its gain and phase margins.
(b) Find the gain and phase margins if the proportional gain is increased to 50.
(c) Discuss the effects on the system performance by the changes in proportional gain examined in (a) and (b).

REFERENCE

[1] R.S. Burns, Advanced Control Engineering, Butterworth Heinemann, Oxford, 2001.

Laplace Transforms Table

	Time Domain, $f(t)$	Laplace Domain, $F(s)$	Notes
1	δt	1	Unit impulse (Dirac delta function)
2	1	$\dfrac{1}{s}$	Unit step
3	t	$\dfrac{1}{s^2}$	Ramp
4	$t^n\ (n = 0, 1, 2, \ldots)$	$\dfrac{n!}{s^{n+1}}$	
5	e^{-at}	$\dfrac{1}{s + a}$	
6	te^{-at}	$\dfrac{1}{(s + a)^2}$	
7	$\cos(\omega t)$	$\dfrac{s}{s^2 + \omega^2}$	
8	$\sin(\omega t)$	$\dfrac{\omega}{s^2 + \omega^2}$	
9	$e^{-at}\cos(\omega t)$	$\dfrac{s + a}{(s + a)^2 + \omega^2}$	
10	$e^{-at}\sin(\omega t)$	$\dfrac{\omega}{(s + a)^2 + \omega^2}$	

Mathematics Background

This appendix presents some of the mathematics background required for the material covered in the main text. It is intended for revision and for use as a reference, and the presentation is therefore simplified. For a more comprehensive description of these topics, please refer to an engineering mathematics textbook.

B.1 COMPLEX NUMBERS

A complex number, z, has real and imaginary parts and can be represented in the complex plane (known as an Argand diagram) as shown below.

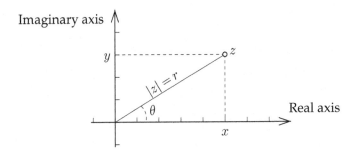

We can represent the complex number z on the Argand diagram in two ways. First, in rectangular coordinates

$$z = x + jy$$

where j is the imaginary operator.[1] Here, x is the real part of z ($x = \text{Re}(z)$) and y is the imaginary part of z ($y = \text{Im}(z)$).

---

[1] As is also done in the text, j (as opposed to i) will be used here for the imaginary operator.

We can also specify this point using polar coordinates, in terms of magnitude r and angle θ to the real axis:

$$z = r\angle\theta.$$

Here, $x = r\cos\theta$ and $y = r\sin\theta$, and the polar form of z becomes

$$z = r(\cos\theta + j\sin\theta).$$

r is the absolute value, or modulus of z, and is equal to

$$|z| = r = \sqrt{x^2 + y^2}.$$

θ is the argument of z, and is equal to

$$\theta = \arg(z) = \arctan\frac{y}{x}.$$

Using the trigonometric identity $e^{jA} = \cos A + j\sin A$ we have:

$$e^{j\theta} = \frac{x}{r} + j\frac{y}{r} \Rightarrow re^{j\theta} = x + jy = z.$$

So we have

$$z = x + jy = re^{j\theta}$$
$$\text{where} \quad r = \sqrt{x^2 + y^2} \quad \text{and} \quad \theta = \arctan\left(\frac{y}{x}\right).$$

B.1.1 Arithmetic with complex numbers

Using rectangular coordinates, the following rules apply for arithmetic operations with two complex numbers $z_1 = x_1 + jy_1$ and $z_2 = x_2 + jy_2$: For addition, we have

$$z_1 + z_2 = (x_1 + jy_1) + (x_2 + jy_2) = (x_1 + x_2) + j(y_1 + y_2).$$

Similarly, for subtraction:

$$z_1 - z_2 = (x_1 + jy_1) - (x_2 + jy_2) = (x_1 - x_2) + j(y_1 - y_2).$$

Multiplication follows normal rules (remember that $j^2 = -1$):

$$z_1 \cdot z_2 = (x_1 + jy_1) \cdot (x_2 + jy_2) = x_1 x_2 + jx_1 y_2 + jy_1 x_2 + j^2 y_1 y_2$$
$$= (x_1 x_2 - y_1 y_2) + j(x_1 y_2 + x_2 y_1).$$

Division of complex numbers can be derived from the multiplication rules:

$$\frac{z_1}{z_2} = \frac{x_1 + jy_1}{x_2 + jy_2} = \frac{x_1x_2 + y_1y_2}{x_2^2 + y_2^2} + j\left(\frac{x_2y_1 - x_1y_2}{x_2^2 + y_2^2}\right).$$

Looking at representation in polar coordinates, some very useful properties for complex numbers can be derived. Our complex numbers are now $z_1 = r_1(\cos\theta_1 + j\sin\theta_1)$ and $z_2 = r_2(\cos\theta_2 + j\sin\theta_2)$. Multiplication in polar form becomes:

$$z_1 \cdot z_2 = r_1 \cdot r_2[\cos(\theta_1)\cos(\theta_2) - \sin(\theta_1)\sin(\theta_2) + j(\sin(\theta_1)\cos(\theta_2)$$

$$+ \cos(\theta_1)\sin(\theta_2))]$$

$$= r_1 \cdot r_2\left[\cos(\theta_1 + \theta_2) + j\sin(\theta_1 + \theta_2)\right].$$

From this, we can derive the following correlations for magnitude and argument of a product of complex numbers:

$$|z_1 \cdot z_2| = |z_1||z_2|$$

$$\arg(z_1 \cdot z_2) = \arg(z_1) + \arg(z_2).$$

Similarly, for division of complex numbers, we have:

$$\left|\frac{z_1}{z_2}\right| = \frac{|z_1|}{|z_2|}$$

$$\arg\left(\frac{z_1}{z_2}\right) = \arg(z_1) - \arg(z_2).$$

B.2 PARTIAL FRACTIONS

Partial fraction expansion is an algebraic technique for simplifying a function-containing polynomial. We want to use it to simplify system transfer function expressions in order to make it easier to find the inverse Laplace transform, preferably by allowing the inverse transfer to be looked up from a table. (See also page 40 for a further example.)

B.2.1 Partial fraction rules

If we have a fraction of the form:

$$\frac{a_0 + a_1s + a_2s^2 + \cdots + a_ns^n}{b_0 + b_1s + b_2s^2 + \cdots + b_ms^m}, \quad \text{e.g.,} \quad \frac{s+2}{s^2 + 4s + 3},$$

it can be expressed in partial fraction form with the following rules:

1. The numerator must be of lower degree than the denominator ($n < m$). If it is not, divide it out first.
2. Factorize the denominator, e.g., $s^2 + 4s + 3 = (s + 1)(s + 3)$.
3. A linear factor $(s + a)$ gives a partial fraction factor $\dfrac{A}{s + a}$.
4. A repeated factor $(s + a)^2$ gives $\dfrac{A}{s + a} + \dfrac{B}{(s + a)^2}$.
5. A cubic factor $(s + a)^3$ gives $\dfrac{A}{s + a} + \dfrac{B}{(s + a)^2} + \dfrac{C}{(s + a)^3}$.
6. A quadratic factor $(s^2 + ps + q)$ gives $\dfrac{As + B}{s^2 + ps + q}$.
7. A repeated quadratic $(s^2 + ps + q)^2$ gives $\dfrac{As + B}{s^2 + ps + q} + \dfrac{Cs + D}{(s^2 + ps + q)^2}$.

B.2.2 Partial fraction example (i)

Find the partial fraction expansion of

$$\frac{6s + 7}{s^2 + 3s + 2}.$$

First, we factorize the denominator, and get

$$\frac{6s + 7}{(s + 1)(s + 2)}.$$

This expression consists of two linear factors of the form $(s + a)$, so using rule 3 we get

$$\frac{6s + 7}{(s + 1)(s + 2)} = \frac{A}{(s + 1)} \quad \frac{B}{(s + 2)}.$$

Multiplying out by the left-hand side denominator and rearranging:

$$6s + 7 = A(s + 2) + B(s + 1)$$

$$6s + 7 = (A + B)s + 2A + B.$$

We see that for this equation to be true, $6 = (A + B)$ (equating coefficients for the s-terms) and $7 = 2A + B$ (equating coefficients for the constant terms). Solving this, we find that $A = 1$ and $B = 5$, so our original transfer function can be written

$$\frac{6s + 7}{s^2 + 3s + 2} = \frac{1}{(s + 1)} \quad \frac{5}{(s + 2)}.$$

B.2.3 Partial fraction example (ii)

Find the partial fraction expansion of

$$\frac{s^2 + 2}{s^3 - 6s^2 + 8s}.$$

Factorize the denominator and put into partial fraction form:

$$\frac{s^2 + 2}{s^3 - 6s^2 + 8s} = \frac{s^2 + 2}{s(s - 2)(s - 4)} = \frac{A}{s} + \frac{B}{(s - 2)} + \frac{C}{(s - 4)}$$

Multiply by $s(s - 2)(s - 4)$ (i.e., the denominator of the LHS):

$$s^2 + 2 = A(s - 2)(s - 4) + Bs(s - 4) + Cs(s - 2)$$

Rearrange for orders of s:

$$s^2 + 2 = A(s - 2)(s - 4) + Bs(s - 4) + Cs(s - 2)$$
$$s^2 + 2 = (A + B + C)s^2 - (6A + 4B + 2C)s + 8A.$$

For the s^0 terms:

$$2 = 8A \Rightarrow A = \frac{1}{4}.$$

For the s^1 terms:

$$0 = -6A - 4B - 2C$$
$$0 = -6/4 - 4B - 2C$$

For the s^2 terms:

$$1 = A + B + C$$
$$1 = 1/4 + B + C$$
$$B = 3/4 - C.$$

Using these two equations, we can find B and C:

$$0 = -6/4 - 12/4 + 2C \Rightarrow C = \frac{9}{4}$$
$$B = 3/4 - 9/4 \Rightarrow B = \frac{-3}{2}.$$

An alternative method, which may be simpler in this case, is to choose some values for s and set in the equation. Starting from

$$s^2 + 2 = A(s - 2)(s - 4) + Bs(s - 4) + Cs(s - 2).$$

Set $s = 0$:

$$0 + 2 = A(0 - 2)(0 - 4) + B \cdot 0(0 - 4) + C \cdot 0(0 - 2)$$

$$\Rightarrow A = \frac{1}{4}$$

Set $s = 2$

$$4 + 2 = A(2 - 2)(2 - 4) + B \cdot 2(2 - 4) + C \cdot 2(2 - 2)$$

$$\Rightarrow B = \frac{-3}{2}$$

Set $s = 4$

$$16 + 2 = A(4 - 2)(4 - 4) + B \cdot 4(4 - 4) + C \cdot 4(4 - 2)$$

$$\Rightarrow C = \frac{9}{4}$$

Hence, our original equation can be written as:

$$\frac{s^2 + 2}{s(s - 2)(s - 4)} = \frac{1}{4s} - \frac{3}{2(s - 2)} + \frac{9}{4(s - 4)}.$$

B.3 DETERMINANTS AND PRINCIPAL MINORS

For the Routh-Hurwitz stability criterion (see page 97) we need to work out the principal minors of the Hurwitz matrix:

$$H_n = \begin{pmatrix} a_1 & a_3 & a_5 & \ldots & \ldots & 0 \\ a_0 & a_2 & a_4 & \ldots & \ldots & 0 \\ 0 & a_1 & a_3 & a_5 & \ldots & 0 \\ 0 & a_0 & a_2 & a_4 & \ldots & 0 \\ 0 & 0 & a_1 & a_3 & \ldots & 0 \\ \vdots & \vdots & \vdots & \vdots & & \vdots \\ 0 & 0 & \ldots & \ldots & \ldots & \end{pmatrix}$$

In practice, we won't calculate determinants larger than 3×3. The determinants can be calculated as follows:

$$\Delta_1 = a_1$$

$$\Delta_2 = \begin{vmatrix} a_1 & a_3 \\ a_0 & a_2 \end{vmatrix} = a_2 a_1 - a_0 a_3$$

$$\Delta_3 = \begin{vmatrix} a_1 & a_3 & a_5 \\ a_0 & a_2 & a_4 \\ 0 & a_1 & a_3 \end{vmatrix} = a_3(a_2 a_1 - a_0 a_3) - a_1(a_1 a_4 - a_0 a_5)$$

Solutions to Questions

This appendix presents the solutions to the questions at the end of the chapters.

Question 1.1

No feedback given.

Question 2.1

Assume that the temperature change is proportional to the temperature difference:

$$\frac{dT_m}{dt} = k(T_a - T_m)$$

$$\frac{dT_m}{dt} + kT_m = kT_a = 0.$$

A solution[1] to this differential equation is

$$T_m(t) = Ae^{-kt}.$$

For $t = 0$:

$$T_m(0) = Ae^{-kt} = 50\,°C \Rightarrow A = 50$$

For $t = 1/k$:

$$T_m(1/k) = 50e^{-1} = 18.39\,°C.$$

Assumptions made:
- The heat capacity of the metal is constant, i.e., the temperature change is proportional to the heat transferred.
- The ambient air remains constant at $0\,°C$.

---

[1] You can find this by trying a solution Ae^{-rt} and substituting it into the differential equation to find r.

Question 2.2

We have

$$C\frac{dv}{dt} = \frac{-v(t)}{R}$$

$$\frac{dv}{dt} = \frac{-v(t)}{CR}$$

A solution to this equation is

$$v(t) = Ae^{-t/CR}.$$

We know from the initial conditions that $v(0) = 5$, hence we find that $A = 5$. For a time $t = CR$, we therefore get

$$v = 5e^{-1} = 1.84\,\text{V}.$$

Question 2.3

We have

$$m = 1\,\text{kg}$$
$$C = 3\,\text{kg/s}$$
$$k = 2\,\text{kg/s}^2$$

The standard form of a second-order system is:

$$\frac{d^2y}{dt^2} + 2\zeta\omega_n\frac{dy}{dt} + \omega_n^2 y = K\omega_n^2 u(t)$$

The mass-spring-damper system in standard form is (see Section 2.3.2):

$$\frac{d^2y}{dt^2} + \frac{C}{m}\frac{dy}{dt} + \frac{k}{m}y = \frac{F_i}{m}.$$

Equating coefficients, we can find the natural frequency ω_n and the damping ratio ζ:

$$\omega_n = \sqrt{\frac{k}{m}} = \sqrt{2} \approx 1.41.$$

$$\zeta = \frac{C}{2\sqrt{km}} = \frac{3}{2\sqrt{2}} \approx 1.06.$$

The damping ratio is just above 1, hence this is close to a *critically damped* system (see Section 2.5.3). The mass will therefore return to its original position quickly and without oscillating.

Question 3.1

We have

$$\frac{d^2y}{dt^2} + \frac{k}{m}y = u(t).$$

Taking the Laplace transform of this expression (see Section 3.2) gives

$$s^2 Y(s) + \frac{k}{m}Y(s) = U(s)$$

$$\frac{Y(s)}{U(s)} = \frac{1}{s^2 + k/m}$$

This is a second-order transfer function with $\zeta = 0$. Hence, $\omega_n^2 = \frac{k}{m}$, and the natural frequency is $\omega_n = \sqrt{\frac{k}{m}}$.

Question 3.2

A standard second-order system has a transfer function (see Section 3.2):

$$G(s) = \frac{K\omega_n}{s^2 + 2\zeta\omega_n s + \omega_n^2}.$$

With $\omega_n = 4$ and $\zeta = 1.25$:

$$G(s) = \frac{4K}{s^2 + 10s + 16}.$$

We can factorize the denominator and use partial fractions to simplify the expression:

$$G(s) = \frac{4K}{(s+2)(s+8)} = \frac{A}{s+2} + \frac{B}{s+8}.$$

Working out A and B (see Appendix B.2) gives the final expression in partial fraction form:

$$G(s) = \frac{4K}{6(s+2)} - \frac{4K}{6(s+8)}.$$

Question 3.3

(a) We have

$$\frac{dh}{dt} + \frac{kh}{A} = \frac{Q_{in}}{A}.$$

Taking Laplace transforms:

$$sH(s) + \frac{k}{A}H(s) = \frac{1}{A}Q_{in}(s)$$

The system transfer function, relating the input to the output, is therefore:

$$\frac{H(s)}{Q_{in}(s)} = \frac{1}{As + k}.$$

(b) For a *unit step input*, the transfer function is $Q_{in} = \frac{1}{s}$ (refer to the Laplace table in Appendix A). The transfer function for the output is then:

$$H(s) = \frac{1}{(As + k)} \cdot \frac{1}{s} = \frac{1/A}{s(s + k/A)}.$$

Using partial fractions, this can be written as

$$\frac{1/A}{s(s + k/A)} = \frac{b}{s} + \frac{c}{s + k/A}$$

Working these out, we get $b = 1/k$ and $c = -1/k$, and the transfer function can be written as

$$H(s) = \frac{1}{k} \cdot \frac{1}{s} - \frac{1/k}{s + k/A}.$$

Taking the inverse Laplace transform, we get

$$h(t) = \frac{1}{k}\left(1 - e^{-k/A \cdot t}\right)$$

For a *unit ramp input*, the input transfer function is $Q_{in} = \frac{1}{s^2}$. The transfer function for the output is:

$$H(s) = \frac{1}{(As + k)} \cdot \frac{1}{s^2} = \frac{1/A}{s^2(s + k/A)}.$$

Using partial fractions, this can be written as

$$\frac{1/A}{s^2(s + k/A)} = \frac{a}{s^2} + \frac{b}{s} + \frac{c}{s + k/A}$$

Working these out, we get $a = 1/k$, $b = -A/k^2$, and $c = A/k^2$, which gives

$$H(s) = \frac{1/l}{s^2} - \frac{A/k^2}{s} + \frac{A/k^2}{s + k/A}.$$

Taking the inverse Laplace transform, the time domain solution is:

$$h(t) = \frac{1}{k}t - \frac{A}{k^2}\left(1 - e^{-k/A \cdot t}\right).$$

Question 3.4

We have

$$\frac{1}{B}\frac{d^2y}{dt^2} + A\frac{dy}{dt} + By(t) = 7u(t).$$

The Laplace transform is

$$\frac{1}{B}s^2Y(s) + AsY(s) + BY(s) = 7U(s)$$

$$\frac{Y(s)}{U(s)} = \frac{7B}{s^2 + ABs + B^2}.$$

For $A = 1$ and $B = 7$:

$$\frac{Y(s)}{U(s)} = \frac{49}{s^2 + 7s + 49}.$$

Comparing the coefficients with the standard form (see Section 3.2), we get $\omega_n = 7$, $\zeta = 0.5$, and $K = 1$. Hence, this is an underdamped system, i.e., will produce an oscillatory response, with a gain of one.

In order for the system not to be oscillatory, $\zeta \geq 1$ (see Section 3.5). Comparing coefficients for the first-order term ($AB = 2\zeta\omega_n$ where $B = \omega_n$), we find that this is satisfied for $A \geq 2$.

Question 3.5

The transfer function for the mass-spring-damper system in standard form is (see Section 3.2.2):

$$G(s) = \frac{\frac{1}{m}}{s^2 + \frac{C}{m}s + \frac{k}{m}}.$$

Comparing with the standard second-order system transfer function, we have $2\zeta\omega_n = C/m$ and $\omega_n^2 = k/m$.

For $\omega_n = 3\,\text{rad/s}$, we get $k = 18\,\text{N/m}$. With $\zeta = 2$, we can find that $C = 24\,\text{Ns/m}$.

This is an overdamped system ($\zeta = 2$), hence the system will not oscillate.

Question 3.6

We can find the characteristics of the responses from coefficients in the transfer functions. Recall the standard form for first- and second-order systems:

$$G_{\text{f.o.}} = \frac{K}{\tau s + 1}$$

$$G_{\text{s.o.}} = \frac{K\omega_n^2}{s^2 + 2\zeta\omega_n s + \omega_n^2}$$

Transfer function 1 is a first-order system with $\tau = 0.2$ and gain $K = 1$. Hence we know that when subjected to a unit step input, the final value will be 1. The response will reach approximately 63% of its final value within one time constant, 0.2 s (see Section 3.4.4). Hence, **plot E** is the correct match.

Transfer function 2 is a second-order system with gain $K = 1/3$. The damping can be estimated by equating the first-order coefficient with the standard form: $1 = 2\zeta\omega_n \Rightarrow \zeta = \dfrac{1}{2\sqrt{30}} = 0.1$. Hence, the system is underdamped and the correct match is **plot C**.

Transfer function 3 is first order, time constant $\tau = 1$, and final value $1 \Rightarrow$ **plot D**.

Transfer function 4 is a second-order system with a gain of 1 and the same damping ratio as transfer function 2, i.e., underdamped. The correct match is **plot B**.

Transfer function 5 is also second order, with a gain of 1/3. The damping ratio can be found similarly as in transfer function 2: $\zeta = \dfrac{10}{2\sqrt{30}} = 0.9$. This is close to critically damped, hence will produce a small, but not significant, overshoot. The correct match is **plot F**.

Transfer function 6 is a first-order system with time constant $\tau = 0.5$ and gain $K = 0.5 \Rightarrow$ **plot A**.

Question 4.1

The transfer function is $G(s) = \dfrac{1}{s + 2}$. Compare with the standard form $\dfrac{1}{\tau s + 1}$, we can rewrite as $G(s) = \dfrac{1/2}{1/2s + 1}$ and we find the time constant to be $\tau = 1/2$.

There are no zeros, as there are no s terms in the numerator. The poles are the solutions to the characteristic equation $s + 2 = 0$, hence there is one pole at -2.

The closed-loop system with unity feedback is (see Table 4.1):

$$G_{\text{cl1}}(s) = \frac{G(s)}{1 + G(s)} = \frac{1}{s + 3}.$$

The new time constant is $\tau = 1/3$, and the new pole is at -3. No zeros have been introduced.

Using a compensator in the feedback loop, we have (again, refer to Table 4.1):

$$G_{\text{cl2}}(s) = \frac{G(s)}{1 + G(s)F(s)}. \tag{C.1}$$

In this case, $F(s)$ is simply a gain of 10, hence we have

$$G_{cl2}(s) = \frac{\frac{1}{s+2}}{1 + \frac{10}{s+2}} = \frac{1}{s+12}. \tag{C.2}$$

The new time constant is $\tau = 1/12$, and there is a new pole at -12. No zeros have been introduced.

The responsiveness of the system improves as the feedback gain increases.

Question 4.2

(a) The open-loop transfer function is

$$G(s) = \frac{1}{(Ls + R)(Js + k)}$$

and the closed-loop transfer function is

$$G_c(s) = \frac{V(s)}{N(s)} = \frac{G(s)}{1 + G(s)} = \frac{1}{(Ls + R)(Js + k) + 1}$$

A PI controller has a transfer function $C(s) = K_p + K_i/s$. The closed-loop block diagram for the controlled system is then

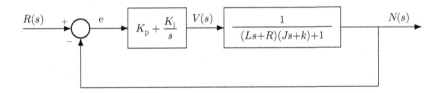

The closed-loop transfer function for the controlled system is (refer to Table 4.1):

$$\frac{N(s)}{R(s)} = \frac{C(s)G_c(s)}{C(s)G_c(s) + 1}.$$

Inserting the transfer functions and simplifying, we get a total system transfer function of

$$\frac{N(s)}{R(s)} = \frac{K_p s + K_i}{LJs^3 + (LK + RJ)s^2 + (Rk + K_p)s + K_i + 1}.$$

Question 4.3

The system block diagram is:

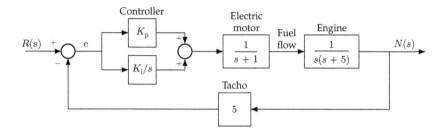

The open-loop (forward path) transfer function is

$$G_o = \frac{K_p + K_i/s}{s(s+1)(s+5)}$$

and the feedback transfer function is simply a gain term: $F(s) = 5$.

Refer to Table 4.1, we can work out the closed-loop transfer function as

$$G_c(s) = \frac{G_o(s)}{1 + G_o(s)F(s)}.$$

Putting in the transfer functions and simplifying, we get

$$G_c(s) = \frac{K_p s + K_i}{s^4 + 6s^3 + 5s^2 + 5K_p s + 5K_i}.$$

At the lower speed of 70 rpm, the closed-loop transfer function is:

$$G_c(s) = \frac{K_p s + K_i}{s^4 + 3s^3 + 2s^2 + 5K_p s + 5K_i}.$$

Question 4.4

In this case we cannot use the block diagram reduction rules directly so we introduce some helping variables in the block diagram, as shown (in lowercase).

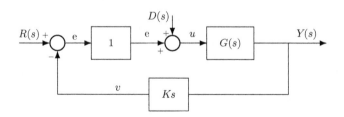

We would like an expression for $\dfrac{Y(s)}{D(s)}$. Working backward from Y (and dropping the "(s)" for clarity) we have:

$$Y = u \cdot G$$

We can substitute for $u = (e + D)$ to get

$$Y = (e + D) \cdot G$$

The signal e is equal to $R - v$, but R is zero so we get

$$Y = (D - v) \cdot G$$

The signal v is simply the output multiplied with the feedback gain ($v = Y \cdot Ks$) and we have

$$Y = (D - YKs) \cdot G$$

This can be rearranged to give:

$$\frac{Y}{D} = \frac{G}{1 + KsG}$$

Inserting for $G(s) = \dfrac{1}{(s + a)(s + b)}$ and simplifying, we get a system transfer function of

$$\frac{Y(s)}{D(s)} = \frac{1}{s^2 + (a + b + K)s + ab}$$

We use the final value theorem to find the steady-state output for a step input in the disturbance $D(s)$:

$$\lim_{t \to \infty} y(t) = \lim_{s \to 0} s \cdot Y(s) = \lim_{s \to 0} s \cdot \frac{1}{s^2 + (a + b + K)s + ab} \cdot \frac{1}{s} = \frac{1}{ab}.$$

As this is a disturbance input, we want the output to be unaffected, i.e., remain zero. We have a steady-state value of $\dfrac{1}{ab}$, hence this system does not reject the disturbance.

Question 5.1

The closed-loop transfer function is

$$\frac{Y(s)}{R(s)} = \frac{\dfrac{\omega_n^2}{s^2 + 2\zeta\omega_n s + \omega_n^2}}{1 + \dfrac{Ks\omega_n^2}{s^2 + 2\zeta\omega_n s + \omega_n^2}}$$

$$= \frac{\omega_n^2}{s^2 + (2\zeta\omega_n + K\omega_n^2)s + \omega_n^2}$$

The Hurwitz matrix and determinants are:

$$H_2 = \begin{pmatrix} a_1 & a_3 \\ a_0 & a_2 \end{pmatrix} = \begin{pmatrix} 2\zeta\omega_n + K\omega_n^2 & 0 \\ 1 & \omega_n^2 \end{pmatrix}$$

$$\Delta_1 = 2\zeta\omega_n + K\omega_n^2 > 0$$

$$\Delta_2 = \begin{vmatrix} 2\zeta\omega_n + K\omega_n^2 & 0 \\ 1 & \omega_n^2 \end{vmatrix} = 2\zeta\omega_n + K\omega_n^2 \cdot \omega_n^2 > 0.$$

Since K, ω_n, and ζ are always positive, these conditions will always be fulfilled, hence the system is always stable.

Question 5.2

The closed-loop transfer function is

$$\frac{Y(s)}{R(s)} = \frac{K\omega_n^2}{s^3 + 2\zeta\omega_n s^2 + \omega_n^2 s + K\omega_n^2} = \frac{K}{s^3 + 2s^2 + s + K}$$

The characteristic equation is $s^3 + 2s^2 + s + K = 0$. We can use the Routh-Hurwitz criterion with the Lienard Chipart rule, so we only need to evaluate Δ_2, which is

$$\Delta_2 = a_1 a_2 - a_0 a_3 = 2\zeta\omega_n \cdot \omega_n^2 - K\omega_n^2$$

(a) Here, we have $\Delta_2 = 2 - K$, hence $K < 2$ for stability.
(b) Generally:

$$\Delta_2 = \omega_n(2\zeta\omega_n - K)$$

ω_n^2 is always positive. For $\Delta_2 > 0$, we need $(2\zeta\omega_n - K) > 0$. Hence, both ζ and ω_n influence the maximum gain that can be used.

Question 5.3

The characteristic equation in standard form is $s^4 + 6s^3 + 5s^2 + 5K_p s + 5K_i = 0$.

The Hurwitz matrix is

$$H_4 = \begin{pmatrix} 6 & 5K_p & 0 & 0 \\ 1 & 5 & 5K_i & 0 \\ 0 & 6 & 5K_p & 0 \\ 0 & 1 & 5 & 5K_i \end{pmatrix}$$

All the coefficients are positive, hence we can use the Lienard Chipart rule. We choose Δ_1 and Δ_3. $\Delta_1 = 6$, hence this is OK for stability.

$$\Delta_3 = \begin{vmatrix} 6 & 5K_p & 0 \\ 1 & 5 & 5K_i \\ 0 & 6 & 5K_p \end{vmatrix} = 30K_p - 36K_i - 5K_p^2.$$

Assuming $K_i = 0$, we have $30 - 5K_p > 0$, hence we must have $K_p < 6$ in *any* case. If $K_i > 0$ then the maximum value of K_p will be limited by the above correlation in order that $\Delta_3 > 0$.

Question 5.4

The forward path transfer function is

$$G = \frac{C(K_p + K_i/s)}{s - 1}$$

The closed-loop transfer function is

$$\frac{Y(s)}{R(s)} = \frac{G}{1 + G} = \frac{CK_p s + CK_i}{s^2 + (CK_p - 1)s + CK_i}$$

For $K_p = K_i = 6$ and $C = 1$, the characteristic equation is $s^2 + 5s + 6 = 0$. For the quadratic equation $as^2 + bs + c = 0$, the solution is $s = \dfrac{-b \pm \sqrt{b^2 - 4ac}}{2a}$, and we get $s_1 = -2$ and $s_2 = -3$, i.e., two real poles.

For $C = 0.5$ we get $s_{1,2} = -1 \pm \sqrt{2}j$, which are the uppermost and lowermost points on the "circle" that the loci create.

Question 5.5

With an open-loop system $G(s) = \dfrac{1}{s(s + 4)(s + 0.1)}$, the closed-loop system is as shown in the figure.

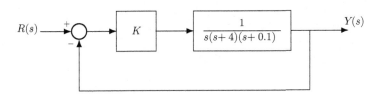

To draw the root locus, we apply the rules:
1. There are three open-loop poles, therefore three loci branches.
2. Any imaginary loci are symmetrical about the real axis.
3. The loci start at the open-loop poles: $0, -0.1, -4$.

4. There are no open-loop zeros, hence all loci will go to infinity.
5. Next, we find the asymptotic angles for the infinite loci. We have the number of zeros, $u = 0$, the number of poles, $v = 3$, hence $r = v - u = 3$, and the angles of the asymptotes are

$$\varphi = (2k + 1)\frac{\pi}{r} \quad \text{for } k = 0, 1, 2:$$

$$k = 0 \implies \varphi = \frac{\pi}{3},$$

$$k = 1 \implies \varphi = \pi,$$

$$k = 2 \implies \varphi = \frac{5\pi}{3}.$$

Hence, the branches go off at angles 60°, 180°, and 300° with the real axis.
6. The asymptotes intersect the real axis at the point

$$a_0 = \frac{\sum_{i=1}^{v} p_i - \sum_{i=1}^{u} z_i}{r} = \frac{((0 - 0.1 - 4))}{3} = -1.37.$$

7. By inspection, we see that the part of the real axis from $-\infty$ to -4 is part of the loci, as is the bit between the two poles at -0.1 and 0. (See this rule explained in the examples in Chapter 5.)
8. There will be a breakaway point for the two poles following the $\pm 60°$-asymptotes. We calculate this by:

$$\sum_{i=1}^{v} \frac{1}{(B_p - p_i)} = \sum_{i=1}^{u} \frac{1}{(B_p - z_i)}$$

$$\frac{1}{B_p + 4} + \frac{1}{B_p + 0.1} + \frac{1}{B_p + 0} = 0.$$

It is clear that it is the two poles close to the imaginary axis that will break away. A first guess for B_p would be the mid-point between these, at -0.05. Trying some values in the expression shows that the breakaway point is in fact very close to -0.05, so we use this value for the sketch.
9. We see from the asymptotes that the loci will cross over into the positive half plane. To find the points on the imaginary axis, we need the closed-loop characteristic equation, which is:

$$1 + \frac{K}{s(s + 4)(s + 0.1)} = 0$$

$$s^3 + 4.1s^2 + 0.4s + K = 0$$

Using Routh-Hurwitz, we get a 3×3 Hurwitz matrix, and since the coefficients are all positive we can use the Lienard Chipart rule. We then only need to consider Δ_2, which is:

$$\Delta_2 = \begin{vmatrix} a_1 & a_3 \\ a_0 & a_2 \end{vmatrix} = \begin{vmatrix} 4.1 & K \\ 1 & 0.4 \end{vmatrix} = 1.64 - K > 0.$$

Hence, for $K = 1.64$, the poles will lie on the imaginary axis. To find the value at these points, we substitute $s = j\omega$ and for K in the characteristic equation:

$$(j\omega)^3 + 4.1(j\omega)^2 + 0.4j\omega + 1.64 = 0$$
$$-j\omega^3 - 4.1\omega^2 + 0.4j\omega + 1.64 = 0.$$

Equating real parts we have $-4.1\omega^2 + 1.64 = 0$, hence $\omega \approx 0.63$.

10. There are no complex open-loop poles or zeros, therefore we don't have to calculate any departure or arrival angles.

The exact root locus plot is shown in the below figure.

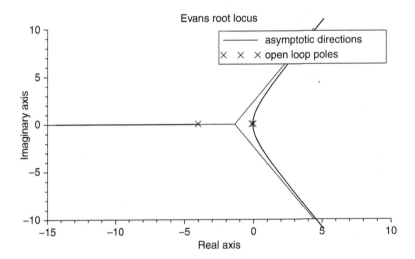

The value of gain that will put the system at the limit of stability was calculated above as $K = 1.64$.

A compensator with a pole at -8 and a zero at -1 has a transfer function $C(s) = \dfrac{s+1}{s+8}$. With the compensator, the closed-loop transfer function is

$$\frac{Y(s)}{R(s)} = \frac{KC(s)G(s)}{1 + KC(s)G(s)}$$

To find the limit of stability, we only need to consider the characteristic equation, which now is $1 + KC(s)G(s) = 0$. Substituting:

$$1 + K \cdot \frac{(s+1)}{(s+8)} \cdot \frac{1}{s(s+4)(s+0.1)} = 0$$

Working this out, we get

$$s^4 + 12.1s^3 + 33.2s^2 + (3.2 + K)s + K = 0$$

Using Routh-Hurwitz, we have:

$$a_0 s^4 + a_1 s^3 + a_2 s^2 + a_3 s + a_4 = 0$$

$$H_4 = \begin{pmatrix} a_1 & a_3 & 0 & 0 \\ a_0 & a_2 & a_4 & 0 \\ 0 & a_1 & a_3 & 0 \\ 0 & a_0 & a_2 & a_4 \end{pmatrix} = \begin{pmatrix} 12.1 & (3.2 + K) & 0 & 0 \\ 1 & 33.2 & K & 0 \\ 0 & 12.1 & (3.2 + K) & 0 \\ 0 & 1 & 33.2 & K \end{pmatrix}$$

Using the Lienard Chipart rule, we consider Δ_1 and Δ_3:

$$\Delta_1 = 12.1 > 0 \ \checkmark$$

$$\Delta_3 = \begin{vmatrix} 12.1 & (3.2 + K) & 0 \\ 1 & 33.2 & K \\ 0 & 12.1 & (3.2 + K) \end{vmatrix}$$

Working out Δ_3 (see Appendix B.3), we get that the condition for stability is $1275.264 + 248.91K - K^2 > 0$. Solving this, we find that the maximum gain that can be used, i.e., the gain at the limit of stability, is $K = 253.9$. Hence, the compensator allows a significantly higher gain to be used compared to the original case.

Question 5.5

The open-loop system comprises a controller G_c and a plant G_p as shown in the figure.

Based on the information given, the system G_p has a transfer function $G_p = \dfrac{1}{s-1}$. The transfer function for a PI controller is $G_c = K_p + \dfrac{K_i}{s} = \dfrac{K_p s + K_i}{s}$. Hence,

$$\frac{Y_o(s)}{U(s)} = \frac{K_p s + K_i}{s(s-1)}$$

For $K_p = 4$ and $K_i = 12$, we see that the numerator is $4s + 12 = 4(s + 3)$, hence there is a zero at $s = -3$. To put the zero at -4, we need $K_p s + K_i = 0$ for $s = -4$. Keeping $K_i = 12$, this is satisfied for $K_p = 3$. The new transfer function for the controller is therefore $\dfrac{3s + 12}{s}$.

In the closed loop, the transfer function is (see Table 4.1):

$$\frac{Y_c(s)}{U(s)} = \frac{G_c G_p}{1 + G_c G_p}$$

Inserting for G_c and G_p and working this out, we get:

$$\frac{Y_c(s)}{U(s)} = \frac{3s + 12}{s^2 + 2s + 12}.$$

Question 6.1

Considering first the $\dfrac{1}{(s + 1)}$ factor. This is a standard first-order $1 + j\omega\tau$ factor, and the corner frequency in this case is 1 (see page 130). The Bode diagram will look like the one shown in Figure 6.8, i.e., a log magnitude of zero up to the corner frequency (1) and then declining at -20 dB/decade. The phase will start at zero, the go down to $-\pi/2$ over two decades around the corner frequency. This corresponds with *Bode diagram A*.

Transfer function 3, $\dfrac{s}{(s + 1)}$ will be the sum of the above factor and the single s factor. The Bode diagram for a single s factor is shown in Figure 6.7b (page 129).

Adding these we get an increasing log magnitude up to the corner frequency (this is the contribution from the s term), while above the corner frequency, the log magnitude stays constant (the -20 and $+20$ dB/decade contributions cancel each other out). Hence the log magnitude plot will look like the one in Figure C.1a.

The phase plot will be the sum of the plots shown in Figure 6.7b and Figure 6.8b, hence will be the same shape as the one for the $1 + j\omega\tau$ factor but shifted $\pi/2$ upward (the contribution from the single s factor). This is shown in Figure C.1b. Overall, this corresponds with *Bode diagram B*.

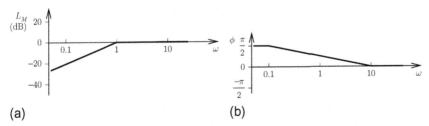

(a)   (b)

FIGURE C.1 Bode plot sketch for $\dfrac{s}{(s + 1)}$. (a) Log magnitude plot and (b) Phase plot.

The $\dfrac{1}{(s+1)(s+100)}$ must be rearranged into standard form (remember that

the standard form for first-order terms are $\dfrac{1}{\tau s + 1}$). This gives

$$\frac{1}{(s+1)(s+100)} = 0.01 \cdot \frac{1}{(s+1)} \cdot \frac{1}{0.01s+1}.$$

With time constants 1 and 0.01, the corner frequencies for the first-order terms are 1 and 100. The log magnitude of the constant term 0.01 is $-40\,\text{dB}$. This gives the individual and combined log magnitude plots as shown in Figure C.2.

The individual phase plot sketches for the first-order terms and the constant gain term (which has zero-phase shift) are shown in Figure C.2b, and the combined plot in Figure C.2d. It can be seen that Figure C.2c and d corresponds with *Bode diagram C*.

For the $\dfrac{s}{(s+1)(s+100)}$, all that is needed is to add the contribution from the single s term in the numerator to the sketches above (see page 129). Figure C.3a shows the log magnitude sketch, which is the one from Figure C.2c but with an added s-only factor increasing by $20\,\text{dB/decade}$. Figure C.3b shows the phase

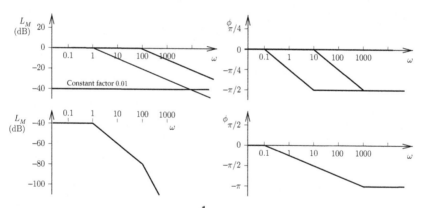

FIGURE C.2 Bode plot sketch for $\dfrac{1}{(s+1)(s+100)}$. (a) Individual log magnitude plots, (b) individual phase plots, (c) combined log magnitude, and (d) combined phase.

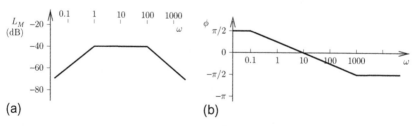

(a) (b)

FIGURE C.3 Bode plot sketch for $\dfrac{s}{(s+1)(s+100)}$. (a) Log magnitude plot and (b) Phase plot.

plot; this is the same as in Figure C.2d but shifted by $\pi/2$ due to the contribution of the s term. This plot corresponds with *Bode diagram C*.

Question 6.2

The open-loop transfer function is

$$G(s) = \frac{2K_p}{s(s+1)(s+5)}$$

(a) For $K_p = 5$:

We rearrange into standard form:

$$G(s) = \frac{10 \cdot 0.2}{s(s+1)(0.2s+1)} = \frac{2}{s(s+1)(0.2s+1)}.$$

(i) Gain term:

$$M(j\omega) = 20\log_{10}|G(j\omega)| = 20\log_{10} 2 = 6.02\,\text{dB}$$
$$\phi(j\omega) = 0$$

(ii) Integral term:

$$M(j\omega) = -20\log_{10}\omega$$

This is a straight line which decays by 20 dB/decade, passing through the point $\omega = 1\,\text{rad/s}$, $M(j\omega) = 0$.

$$\phi(j\omega) = -90° \quad \text{(independent of frequency)}$$

(iii) First-order term:

$$M(j\omega) = 20\log_{10}|G(j\omega)| = 20\log_{10}\left|\frac{1}{1+j\omega t}\right| = -20\log_{10}\sqrt{1+\omega^2\tau^2}$$

At low frequencies ($\omega \ll \tau$):

$$M(j\omega) = -20\log_{10} 1 = 0\,\text{dB}$$
$$\phi(j\omega) = -\tan^{-1}\omega\tau \approx 0°.$$

At high frequencies, find the corner frequency $\omega = \dfrac{1}{\tau}$. The log magnitude decays by 20 dB/decade above this.

The phase is:

$$\phi(j\omega) = -90°.$$

The corner frequencies for the two first-order terms are:

$$\frac{1}{s+1} = \frac{1}{j\omega + 1} \Rightarrow \omega_1 = \frac{1}{\tau_1} = 1 \text{ rad/s.}$$

$$\frac{1}{0.2s+1} = \frac{1}{0.2j\omega + 1} \Rightarrow \omega_2 = \frac{1}{\tau_2} = 5 \text{ rad/s.}$$

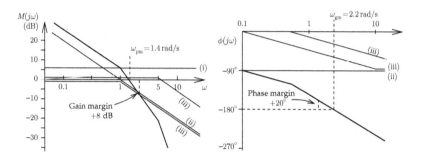

From the Bode plot, the gain margin can be read off as $+8$ dB and the phase margin is $+20°$. (Plotting on logarithmic graph paper allows more accurate readings; the sketches shown are for guidance only.)

(b) For $K_p = 50$:

$$G(s) = \frac{100 \cdot 0.2}{s(s+1)(0.2s+1)} = \frac{20}{s(s+1)(0.2s+1)}.$$

(i) Gain term:

$$M(j\omega) = 20\log_{10} 20 = 26.02 \text{ dB}$$
$$\phi(j\omega) = 0$$

All other terms are the same as in question (a).

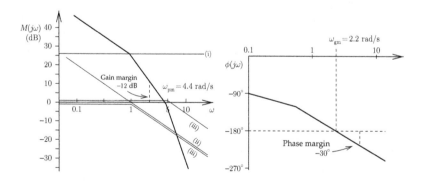

From the Bode plot, the gain margin is -12 dB and the phase margin is $-30°$.

(c) Discussion

The increase in proportional gain from $K_p = 5$ to $K_p = 50$ effectively results in a shift in the 0 dB axis by 20 dB. The system is stable for $K_p = 5$ and unstable for $K_p = 50$.

When $K_p = 5$, the system gain may be increased by 8 dB before it becomes unstable.

To obtain satisfactory performance, the phase margin must be increased by 30-60° which can be achieved by reducing K_p further; however, this would result in a large steady-state error. Alternative compensation can be used to improve the controlled performance of the system.

Index

Note: Page numbers followed by f indicate figures and t indicate tables.